KV-638-326

SCOTLAND'S ENVIRONMENT

during the last 30,000 years

Frontispiece:
Hummocky moraine—Glen Torridon (Inst. of Geol. Sci. photo.)
Published by permission of the Director, N.E.R.C. copyright.

SCOTLAND'S ENVIRONMENT during the last 30,000 years

R. J. PRICE
Reader in Geography,
University of Glasgow

1983

SCOTTISH ACADEMIC PRESS
EDINBURGH

Published by
Scottish Academic Press Ltd,
33 Montgomery Street
Edinburgh EH7 5JX

SBN 7073 0325 7

Design, typography and layouts
by T. L. Jenkins

Printed by
The Universities Press (Belfast) Ltd., Northern Ireland.

CONTENTS

Landsat Band 7, mosaic of Scotland. Produced by digital image processing by the Royal Aircraft Establishment, Farnborough and published with its permission.

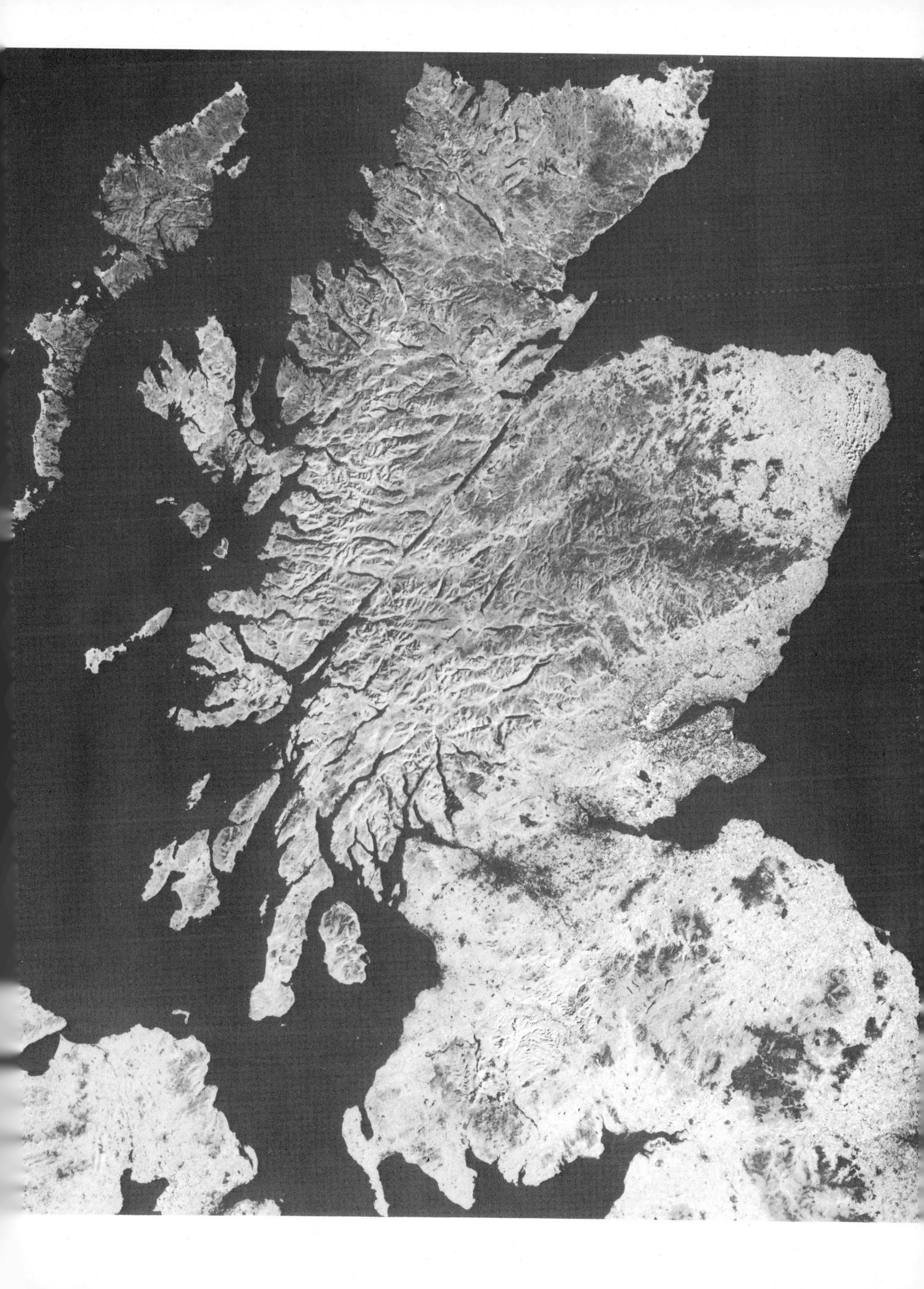

LIST OF TABLES

PREFACE AND ACKNOWLEDGEMENTS

In his editorial in the first volume of the new journal *Quaternary Science Reviews* (1982), D. Q. Bowen states, "All the Quaternary sciences have recently undergone a vast data explosion reaching levels of activity scarcely imaginable a mere decade ago. To a large extent this has resulted from the fashioning of powerful linkages between the sciences in this essentially multi-disciplinary and interdisciplinary field". One hundred and forty years ago the implications of the Glacial Theory (the existence in the recent geological past of great ice sheets and glaciers) began to be realised by geologists interpreting the superficial deposits and their included fossils in Scotland. Within thirty years Archibald and James Geikie and T. F. Jamieson had produced a framework of environmental changes in Scotland into which the sedimentary, palaeontological and geomorphological evidence collected by them and their co-workers could be fitted. This framework remained largely unaltered for nearly one hundred years and apart from a re-evaluation of the chronology with the application of radiometric dating techniques the original framework of environmental changes which occurred during and since the last glaciation still remains valid. However, there has been a great explosion in the publication of papers relating to the Quaternary Period in Scotland during the last twenty years and there have been significant developments in techniques of data collection and analysis.

This book attempts a synthesis of the results obtained by a variety of specialists whose work contributes to our understanding of palaeoenvironments. Inevitably I have become involved in fields of research with which I have little familiarity. While recognising the excellent contributions of specialists—palynologists, palaeontologists, sedimentary geologists, radiometric dating experts, geomorphologists and archaeologists—I offer my apologies to them, if, in any generalisations I have made, their work is misrepresented. By frequent consultation with colleagues at the University of Glasgow and at other institutions, I hope that some of the pitfalls of generalisation have been avoided. I am particularly indebted to Dr. G. Jardine (geologist) who read the entire manuscript and provided many useful suggestions. Dr. J. Dickson (botanist), Dr. D. Smith (geomorphologist), Dr. D. Q. Bowen (geomorphologist) and Professor L. Alcock, Mr. A. Morrison, Dr. E. Slater (archaeologists), all read and commented on sections of the manuscript. To all these people I am not only grateful for their critical comments and suggestions but also for the many enjoyable hours of informal discussion of Quaternary matters.

During the past twenty years I have been privileged to undertake research not only in Scotland but in Alaska, Iceland, Sweden and the Yukon and to visit many research institutes in the United States of America, Canada and in various countries in western Europe. To the Court of the University of Glasgow, the Carnegie Trust for the Universities of Scotland, the Royal Society and the Natural Environment Research Council who have provided financial support for research and travel, I am most grateful.

The preparation of this book has been greatly assisted by the following persons: Mrs D. Briggs typed the manuscript; Miss Yvonne Wilson drafted the maps and diagrams; Mrs H. Davies produced the index. Many authors kindly gave their permission for the reproduction of their maps and diagrams and their names are duly recorded in the figure captions. Mr. Douglas Grant of the Scottish Academic Press has been a constant source of encouragement and helpful advice.

R. J. PRICE
Glasgow 1983

1

INTRODUCTION

The word 'environment' is used in many different ways and in recent years it has taken on an emotive quality. In the title of this book, 'environment' is used to refer to a set of objects and conditions which occur at or near to the earth's surface. Much of this book is concerned with the natural environment, that is, the totality of natural attributes in a given area and over a specified period of time and unaffected by any form of human intervention. The location is Scotland and the time period is 30,000 years. It is implicit in the choice of area and time period that there will be variations in environmental conditions within the chosen area and over the chosen time period. While the prime aim of the book is to identify natural variations in environmental conditions both in space and over time, the fact that Scotland has contained a human population for at least the last 8,000 years introduces the possibility of human modification of the natural environment.

The study of the environment involves a wide variety of natural sciences—geology, climatology, pedology, hydrology, botany, zoology and oceanography. The discipline of physical geography attempts to integrate these scientific investigations into a coherent description and explanation of environmental conditions at selected spatial and temporal scales. The discipline of ecology is concerned with the study of the relationships between living organisms and the physical environments in which they live, while palaeoecology is concerned with the relationships between dead organisms and their former environments. In practice, however, palaeoecology is dependent upon both geological and biological evidence to reconstruct former environments and it is extremely difficult to deduce the relationships between the organisms and their environment in the past, if the evidence of the organisms has already been used to reconstruct the environment.

The identification of changes in environmental conditions over a time period of 30,000 years requires heavy dependence on analogy and deduction. The study of modern physical environments allows the identification of the processes responsible for the character of modern sediments and the environmental parameters required by the various species of plant and animal life. On the assumption that the same processes and environmental parameters were operative in the past, sediments and fossils unrelated to present environmental conditions in a given area can be used to reconstruct former environments. A knowledge of modern sedimentary processes and ecological controls in a wide variety of natural environments is necessary before palaeoenvironments can be reconstructed (Birks & Birks 1980).

The study of the environmental conditions existing in a given area is highly dependent upon the spatial and temporal scales at which the study is undertaken. It

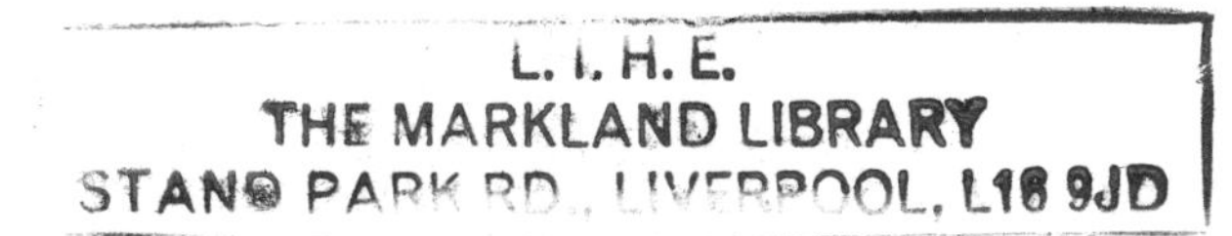

is common practice to build up regional patterns of environmental conditions from data obtained at individual sites. These may be individual exposures of solid or unconsolidated rock, a borehole in lacustrine sediments or an individual meteorological station. Depending on the size of the area studied and the density of data points, the accuracy of the areal generalisations will vary. Since the various processes operating to create a given environment operate on different time-scales the linkages between different parts of the environmental system may appear to change depending upon the time-scale adopted in any particular investigation.

The complexities introduced by spatial and temporal variations in environmental conditions are considerable. It is extremely difficult to give an accurate description of Scotland's environment involving the entire country on a particular day at a particular time. Although landforms which develop over thousands or even millions of years may be regarded as stable elements of the environment at a particular time, the climatic conditions are changing by the hour or even by the minute. This leads to the necessity of producing a description of average conditions for an unspecified time period usually referred to as 'the present'. Geological, topographical, botanical, zoological and climatological maps can be constructed representing the distribution of those elements of the environment as they have been deduced by fieldwork (data collection) over the past 100 years. Such maps immediately reveal the interplay between the various systems through such controls as parent material, altitude, aspect and rainfall. The complexities of the real world are generalised into statements about variations in environmental conditions in different parts of Scotland. It is by comparison with these generalised statements about present environmental conditions that the extent of environmental changes in the past can be assessed.

The definition of the term 'Scotland' also poses problems. The present political entity consists essentially of the land area above mean high water mark along the coast of the mainland and islands north of the border with England. In certain contexts this political boundary is extended beyond the present coastline to include the surrounding seas. From the point of view of the study of environmental change it is more reasonable to define Scotland as not only the mainland and islands but also the waters and the underlying rocks of the continental shelf.

The data base for making generalisations about the present environment of Scotland is far greater than the data base for reconstructing palaeoenvironments. The accuracy of reconstructions is probably inversely proportional to the age of the period for which any reconstruction is attempted. Our understanding of environmental conditions in Scotland over the period 1·6 million to 30,000 years ago is mainly dependent upon data from sediments and fossils obtained in England, on the European continent or the floor of the North Atlantic ocean. Broad generalisations are possible but internal variations within Scotland over that time period are simply not known. The reconstruction of Scottish palaeoenvironments during the last 30,000 years is possible because of the discovery of sediments and fossils in Scotland which can be dated and which reveal something of the environmental conditions associated with those sediments and the plant and animal remains included in them. The density of data points for the period prior to the last glaciation is very low (a few dozen sites) and even the number of sites which provide information about Lateglacial and Postglacial environments only amounts to a few hundred. As will be demonstrated in the next section on the present physical environment of Scotland there are great internal variations in all aspects of the physical environment. Statements about palaeoenvironments based on a limited data base must therefore remain only approximations.

THE PRESENT PHYSICAL ENVIRONMENT

Scotland is located between 55°N and 60°N latitude on the north-western margin of the continent of Europe. Although it is a relatively small country it is geologically and geomorphologically very complex and exhibits a great diversity in terrain characteristics. In general terms the country experiences climatic conditions which are surprisingly mild in relation to its latitudinal position but because of the effects of altitude, aspect and exposure, local climatic variability is considerable.

There are numerous detailed summary statements describing most aspects of Scotland's present environment (O'Dell & Walton 1962, McVean & Ratcliffe 1962, Burnett 1964, Darling & Boyd 1969, Sissons 1967, 1976, Craig 1970, Tivy 1973). There is no single overview of the country's climate or soils but the former is dealt with in many publications produced by H.M.S.O. Meteorological Office and the latter in publications of the Soil Survey of Scotland.

No attempt is to be made here to provide a comprehensive description of all aspects of Scotland's present environment. Such a task would require a separate volume. It is necessary, however, to identify the major regional variations in geology, landforms, climate, soil and vegetation in order that the evolutionary changes of the last 30,000 years to be discussed in subsequent chapters may be put in context.

Three geological/geomorphological regions can be identified in Scotland—the Southern Uplands, the Central Lowlands and the Highlands and Islands. Sissons (1967a, p. 3), has pointed out that there is ". . . a striking correlation in many areas between rock type and scenery, a correlation that applied both to altitude and form of the ground."

The Southern Uplands consist of a series of smooth rounded hills with only a small percentage of the area attaining altitudes in excess of 600 m and only very small areas over 800 m. Much of this region is underlain by steeply-inclinded beds of Ordovician and Silurian greywackes and mudstones. There is a strong north-east to south-west structural trend which is quite strongly reflected in ridge and valley orientation. In the south-west of the region granite masses coincide with the high ground of Cairnsmore of Carsphairn, Cairnsmore of Fleet and Criffell. The Loch Doon basin is also underlain by granite while the surrounding metamorphic rocks coincide with a rim of high ground.

The Central Lowlands are bounded both to the north and south by major fault zones. This region is either underlain by Old Red Sandstone and Carboniferous sediments, which generally are associated with low ground, or intrusive and extrusive igneous rocks which are strongly correlated with areas of high ground. There really is no satisfactory single morphological term which can be used to label the region. Quite significant areas of the "Central Lowlands" consist of Carboniferous lavas which are bounded by steep fault-line scarps with extensive uplands over 400 m above sea level (e.g. the Kilpatrick, Campsie and Ochil Hills). Isolated volcanic necks have produced steep-sided hills which add to the complexity of the terrain (e.g., Dumbarton Rock, Arthurs Seat, North Berwick Law). Much of the lower part of the Clyde Basin is underlain by Carboniferous shales and weak sandstone (massive limestones are rare), while large areas of the Lower Forth and Tay Basins are underlain by Old Red Sandstone sediments. Only along the Highland border do the Old Red Sandstone rocks form conspicuous hills.

The Highlands and Islands constitute a very complex region capable of subdivision into several distinct sub-regions on the basis of their landform characteristics (Price 1976). The Grampians (including the Cairngorms) and the north-central Highlands consist of a series of ridges, valleys and small plateau surfaces developed on ancient metamorphic rocks and granitic intrusions. There are extensive areas over 800 m while many summits attain altitudes in excess of 1,000 m (e.g. Ben Nevis, 1,347 m; Ben Macdhui, 1,311 m). It is not the extent of high altitude ground which dominates these areas but the amount of dissection and dominance of steep slopes. The extreme north-west of the mainland (north-west of the Moine Thrust) and many of the Outer Hebridean Islands are underlain by the oldest rocks in Scotland. Much of the area has the basement Lewisian gneiss outcropping at the surface but in some locations the gneiss is overlain by Torridonian sandstone which produces distinctive hill-masses. In the north-east of Scotland, in Caithness, Orkney and around the shores of the Moray Firth, areas of relatively low ground are underlain by Old Red Sandstone sediments. In the Inner Hebridean Islands of Skye, Rhum, Eigg, Mull and Arran some very distinctive terrain is associated with Tertiary volcanic rocks.

The above brief description of the geology and landforms of Scotland can perhaps best be summarised in terms of the distribution patterns of absolute altitude and relative relief. South of the Highland Boundary Fault there are large areas below 200 m in which relative relief values rarely exceed 100 m and slopes are generally of low angle. The core of the Southern Uplands and the individual hill masses in the Central Lowlands contain considerable areas between 400 m and 800 m in which relative relief values range between 100 m and 400 m. These areas contain many steep slopes. With the exception of north-east Scotland and the Outer Hebrides the Highlands and Islands are characterised by large areas between 600 m and 1,000 m with relative relief values ranging between 400 m and 600 m and, because of the highly dissected nature of the terrain, steep slopes are dominant.

The climate of Scotland is surprisingly poorly documented. This is particularly true of the upland areas (over 400 m) where there are only six meteorological stations with records for three years or more (Harding 1978). Figure 1.1 shows that in January much of the country at sea-level has a mean temperature of between 3°C and 4°C with much of the upland areas being two or three degrees cooler. On the summit of Ben Nevis (1,347 m) mean monthly temperatures are at, or below, 0°C for eight months of the year. Sea surface temperatures on the west coast of Scotland in January are about 1°C warmer (8°C) than those on the east coast (7°C). In summer most of Scotland has a mean sea-level temperature of between 13°C and 15°C with only a small area in the Central Lowlands having a mean July temperature of over 15°C. Again the actual surface temperatures in July are 2°C–5°C cooler in the upland areas. Sea surface temperatures in July are 1°C cooler on the west coast (12°C) than they are on the east coast (13°C).

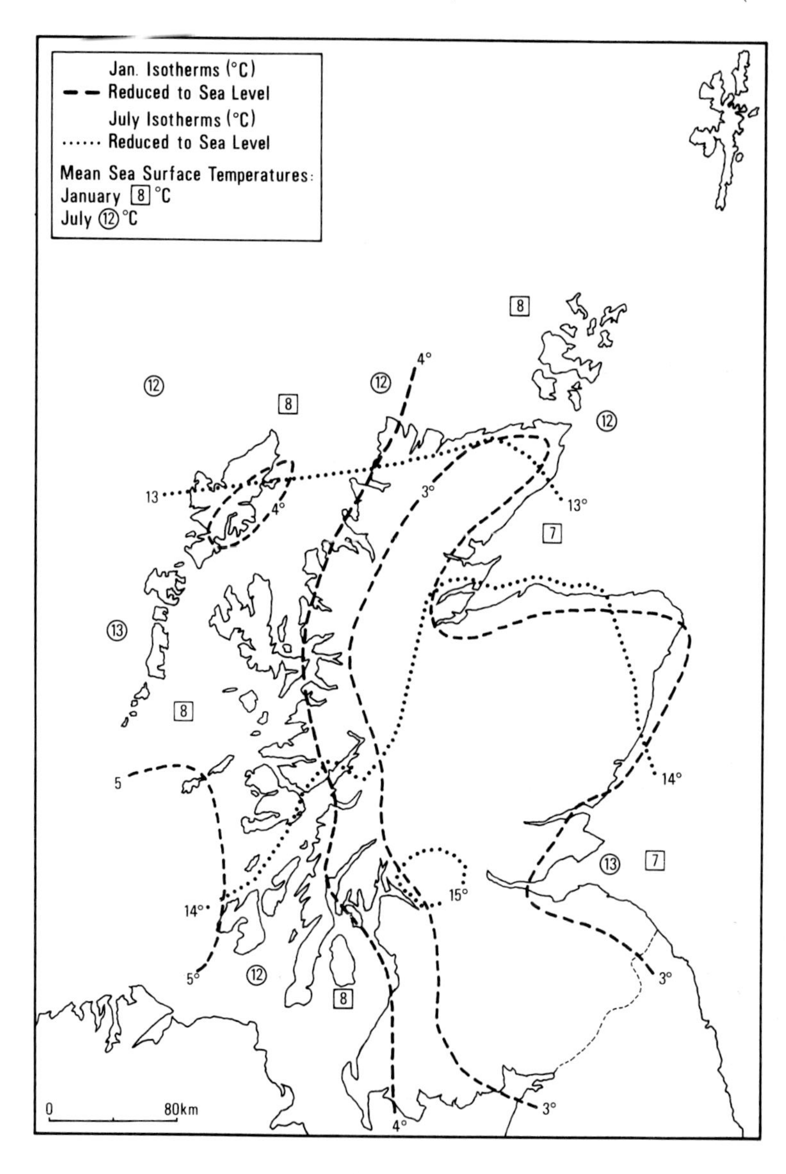

Fig. 1.1

Mean January and July temperatures.

Compared with other land masses at similar latitudes in the northern hemisphere, Scotland enjoys a very mild climate; Scotland lies down-stream and down wind of what Mackinder (1915) described as the world's greatest "winter gulf of warmth." It is the presence of the relatively warm waters of the Gulf Stream to the west of Scotland and the passage of air masses across those warm waters that provides Scotland with such a moderate climate. However, because of the great topographic variability of the country there are marked contrasts between lowland and upland sites. Temperatures diminish with altitude in the moist Atlantic air at a lapse rate rather faster than 0·65°C per 100 m and wind speed, cloudiness and precipitation also increase rapidly.

It is not surprising that there is a high correlation between average annual precipitation totals and altitude and that precipitation totals generally decrease from west to east (Fig. 1.2). Virtually all areas above 400 m in altitude receive more

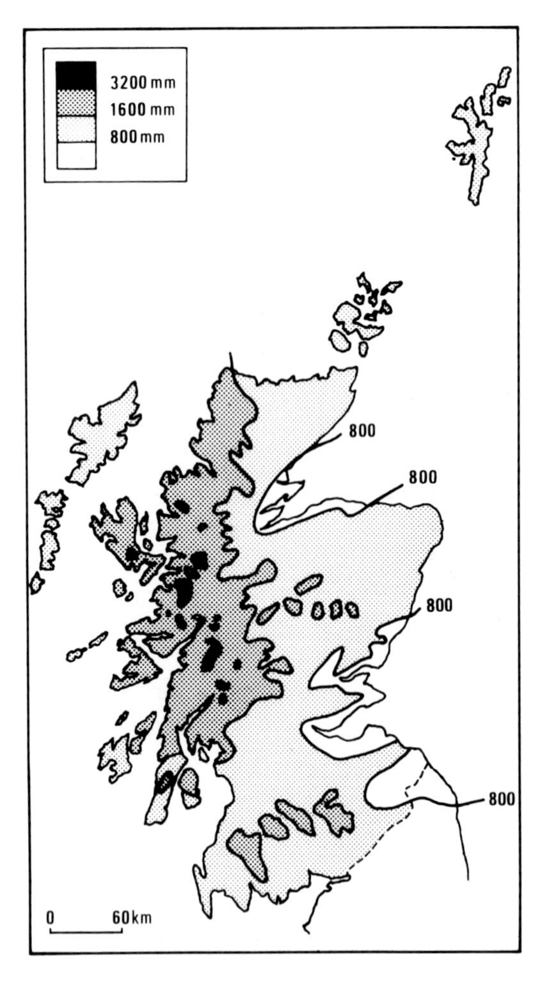

Fig. 1.2.

Mean annual precipitation.

than 1,000 mm of precipitation per year. Only the lowlands around the Moray Firth, of eastern Buchan and East Lothian record totals lower than 600 mm. Large areas in the western mountains receive more than 2,500 mm with over 4,000 mm being recorded inland from Loch Nevis. Scotland has a potential evapotranspiration rate of only about 400 mm so that over large areas of the country soils are waterlogged for long periods and anaerobic conditions commonly lead to peat formation.

Scotland's soils are generally young and immature in that most of them have developed during the last 10,000 years. Soils began to form soon after the land emerged from beneath the last ice sheet (circa 14,000 bp) but many were destroyed by the return of cold conditions during the Loch Lomond Stadial (circa 11,000–10,000 bp). Young soils are greatly affected by their parent materials and the many lithological types which outcrop at the surface and which can vary greatly over short distances tend to produce a wide variety of physical and chemical attributes in the soils. The situation is further complicated by the fact that many soils have developed on glacial drifts each with a different admixture of rock types. Local climatic variations, particularly vertical zonation of precipitation and temperature also lead to different environments for soil forming processes. Much of Scotland was covered by forests within the first few thousand years of the Postglacial period. In the south of the country the birch-oak forests tended to produce brown earths with some tendency towards podzolisation. In the north and west of Scotland the pine/birch forests developed iron and humus podzols. The subsequent removal of much of this forest cover (see below) has led to either the expansion of peat cover or to soil erosion. On the coastal lowlands the presence of raised beach sands and gravels, raised estuarine silts and clays, fluvioglacial sands and gravels and blown sand, all provide distinctive and locally important soil characteristics.

The present vegetation and wild animal population of Scotland has been greatly influenced by human activity over the last 5,000 years. Before the human population began to develop agricultural and industrial techniques the flora and fauna of Scotland was much richer. The present vegetation consists of moorlands (66 per cent of total land area), woodland (nine per cent) and improved land. The moorlands consists of heather, grasses and sedges and they occur above the limit of cultivation and systematic agricultural improvement which lies at about 200 m in the west, 400 m in the south-east and at sea-level in the north-west of Scotland. They contain relatively few plant species because of the harsh physical environment and the effects of uncontrolled grazing of game and domestic animals and of burning (muirburn). Most of the present moorland area was formerly covered by forests up to altitudes between 600 m and 800 m in the east and 300 m in the north-west of Scotland. Closely associated with the moorlands is the distribution of peat. Peat is defined as a surface organic layer not less than 30 cm thick and containing 80 per cent organic matter (dry weight). Approximately ten per cent of the land area of Scotland is covered by peat.

The present woodlands of Scotland, including areas of recent afforestation, cover 8·5 per cent of the total land area of Scotland. There are probably only a few very small forested areas which have not in any way been modified by man or domesticated animals. There are only three indigenous species of conifers, Scots pine (*Pinus sylvestris*), yew (*Taxus baccata*) and juniper (*Juniperus communis*). Indigenous broadleaved species number less than forty, of which the oak (*Quercus* spp), birch (*Betula* spp), alder (*Alnus glutinosa*), hazel (*Corylus avellana*), willow (*Salix* spp), rowan (*Sorbus aucuparia*) and elm (*Ulmus glabra*) are the most important.

The wild animal population of Scotland has changed greatly over the period of forest contraction and the growth of human population. The red deer (*Cervus elaphus scoticus*) is Scotland's largest indigenous mammal. With the exception of a small area in Galloway, red deer are confined to the Highlands and Islands and in 1968 the herd was estimated to have a total population in excess of 180,000. While the red deer is to be found on treeless slopes of the high ground, the roe deer (a much smaller population) is associated with woodland habitats. Apart from the fox, otter, wild cat and pine martin, many of the mammals which returned to Scotland at the beginning of the Postglacial period have died out either directly as a result of hunting or indirectly as a result of human modification of their habitats. The giant Irish Elk, the Great Elk, the Reindeer, the northern Lynx, the Brown Bear, the Wild Boar and the Wolf all once lived in Scotland.

PALAEOENVIRONMENTS—THE NATURE OF THE EVIDENCE

Consolidated and unconsolidated sediments and their included fossils are undoubtedly the most important source of information about past environments. The sediments themselves can reveal by their structures, grain size and shape and mineralogical content, the environment of deposition, the type of depositing medium and the source areas of the sediments involved. It is possible to determine whether sediments were laid down in a marine environment, within a river channel or in a lake. It is equally possible to determine whether a sediment was deposited by a river, by a glacier or by the wind. Sediments, therefore, can reveal a great deal about the environmental conditions existing at the time of their deposition but if they contain fragments of the plant and animal life which existed in that environment and the genus and or species can be identified then, by analogy with the environments currently occupied by such species, a much more accurate picture of the former environment can be constructed.

The Quaternary Period—approximately the last two million years—has been characterised by numerous and significant climatic changes in all parts of the world. The expansion of the great ice sheets to cover about 30 per cent of the total land area on several occasions in the Quaternary was the most dramatic expression of these climatic changes. The expansion and contraction of the ice sheets was also accompanied by the fluctuations in relative sea-level of approximately 100 m. Differences in mean annual temperatures between alternating warm (interglacial) and cold (glacial) phases ranged from 5°C in maritime locations to greater than 10°C in the interior of continents. Environmental changes of such magnitude resulted in the deposition of sediments of widely differing character and caused rapid migration and sometimes extinction of plant and animal communities.

Two million years is a very short period of time in earth history and the rapidity of environmental change during the Quaternary period makes it both a fascinating and often frustrating period to study. Once the theory of repeated glaciation in the recent past had been generally accepted after 1860 AD a great deal of stratigraphical, morphological and palaeontological evidence was collected which demonstrated that alternating periods of cold and warm environments have existed in middle and higher latitudes during the Quaternary period. A fairly simplistic model of four glacial phases separated by interglacial phases was suggested and much of the field evidence was fitted into this model without any real understanding of the chronology of these phases. During the last 30 years two major developments have taken place which have allowed great progress in our understanding of the Quaternary. Firstly, a timescale has been constructed based on various radiometric and geomagnetic techniques (radiocarbon, uranium series, potassium argon and geomagnetic reversals). Secondly, exploration of the ocean floors has revealed much more detailed palaeontological and geochemical records of climatic change than is available from sediments accumulated on the continental land areas. The history of the application of these various techniques and the discoveries that they have produced has been excellently presented by John & Katherine Imbrie in their book *Ice Ages—Solving the Mystery*. There is, of course, a range of specialist books and monographs on techniques and processes available to students of the Quaternary, which are useful in interpreting the character of the environmental changes which occurred (Flint 1971, Goldthwait 1971, Price 1973, Sugden & John 1976, West 1977, Bowen 1978, Birks & Birks 1980). No attempt will be made to discuss these techniques and processes in detail but by means of a simplified model of a typical sequence of sediments, fossils and landforms likely to be produced by a series of environmental changes typical of an interglacial-glacial-interglacial cycle, the different types of evidence to be used in the remainder of this book can be illustrated.

There are five major environmental zones in which the evidence for environmental changes can be collected (Fig. 1.3). Within each of these there are distinctive sediments, fossils and morphology which can be used as indicators of environmental changes related to one cycle of warm-cold-warm.

The mountain zone

Traditionally, the mountain areas of the world have been regarded as those where the most dramatic expressions of a period of glaciation are to be found. It was in the European Alps and in the mountains of Scandinavia and Scotland that the idea of the former expansion of glaciers and ice sheets developed. It was by comparison of the deposits and landforms near existing mountain glaciers with similar deposits in mountain areas which did not contain glaciers that the 'glacial theory' became established. A typical sedimentary sequence in a mountain area (Fig. 1.3a) might consist of bedrock overlain by angular rock fragments produced by freeze-thaw processes at high altitude during the cooling phase of an interglacial. These frost-shattered rocks would move downslope and would be buried by sand and rounded gravel deposits produced during the early phases of glacier expansion.

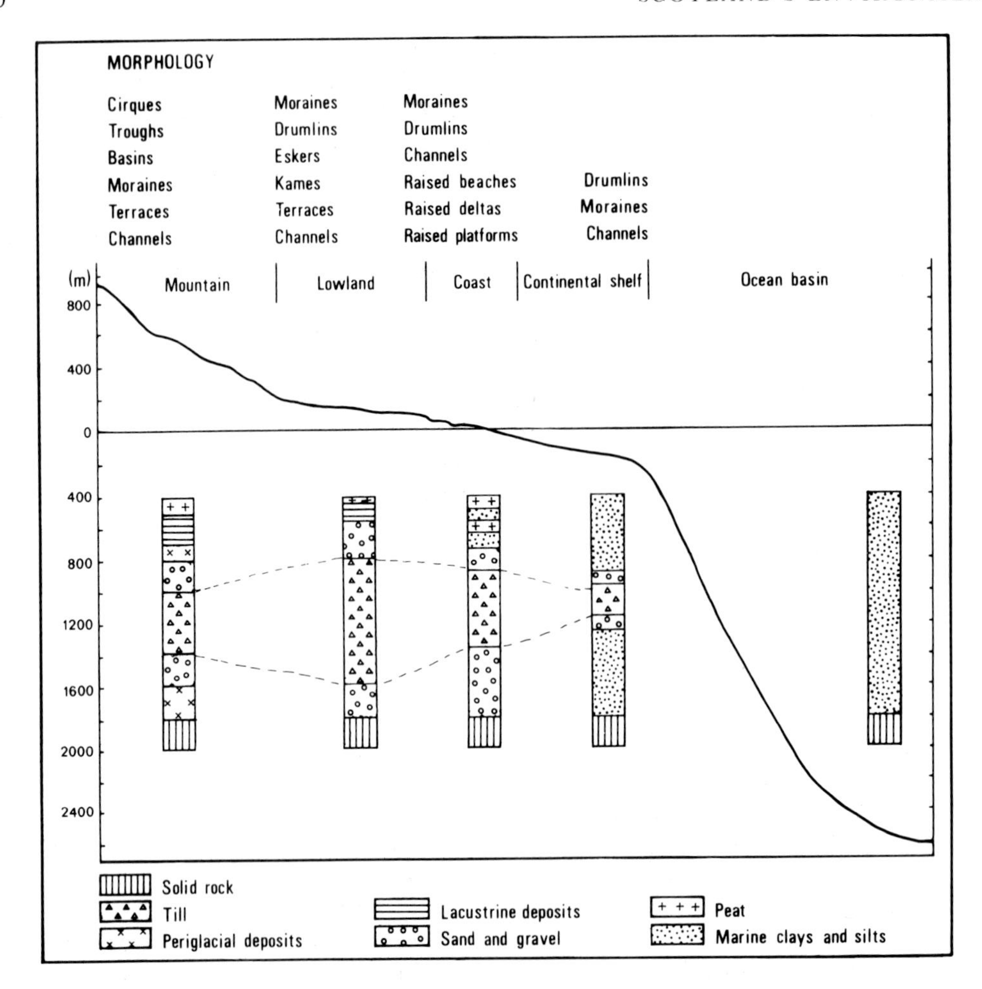

Fig. 1.3. Environmental zones—morphology and deposits.

These outwash deposits (valley-train or valley-sandur) would in turn be overlain by till—the deposits of the advancing valley glacier—consisting of angular fragments of rock in a silty-clay matrix and exhibiting little evidence of sorting or stratification. During the retreat of the valley glacier, the release of large quantities of melt-water would result in the burial of the glacial till by thick accumulations of stratified sand and gravel (outwash material). Depending upon local conditions either the outwash deposits may be dissected to produce river terraces or lakes may develop where lacustrine sediments would accumulate which in turn, on the infilling of the lake basins, may be covered with a layer of peat. Because of the high energy potential of mountain environments glacial erosion tends to be severe. The morphological evidence of a period of glaciation is therefore dominant and major glacial troughs, corries, breached-watersheds, hanging valleys and rock basins are the main expressions of the glacial event. Any evidence of the preceding interglacial period or even of earlier periods of glacial activity may be entirely removed by the last glacial event. For these reasons the mountain zone is not often a major source of stratigraphical and palaeontological information about environmental changes which took place prior to the last glacial event. The outwash gravels on top of the glacial till are usually dissected to produce terraces. Within and on top of these terrace gravels there are often lacustrine sediments and peat layers which contain both macro- and micro-animal and vegetational remains. It was the study of the fossils in these upper sediments which first provided the evidence of the dramatic climatic changes which are associated with the change from a glacial to an interglacial environment. In particular, the detailed study of pollen grains (palynology) allowed the reconstruction of past vegetation cover (Birks and Birks 1980).

The lowland zone

In general terms, areas of small relative relief are more likely to contain sedimentary records of Quaternary environmental changes. The classical 'drift' sequences of the northern European Plain and of the Midwest of the United States bear testimony to this. The sedimentary sequence (Fig. 1.3b) is likely to consist of fluvial, lacustrine or aeolian deposits of an interglacial environment overlain by glacial till and succeeded again by the interglacial sequence. If these deposits contain mammalian fossils, coleoptera, molluscs, pollen and macro vegetation fragments then much more information can be gained about environmental conditions as each part of the sedimentary sequence accumulated. The interglacial sediments are much more likely to be fossiliferous than the sediments deposited during the glacial interval as temperature conditions during the glacial period would hardly be conducive to the existence of many life forms. There are, of course, many problems associated with the interpretation of environmental conditions that existed over a wide area from the fossil fauna and flora derived from a few sites. The fact that many fossils can be washed or blown to their final resting place and that some species may be completely unrepresented because of their failure to survive transportation and deposition make palaeoenvironmental reconstruction from such evidence rather hazardous. One of the most useful indicators of former climatic conditions in terrestrial environments are Coleoptera (beetle) assemblages. Their robust skeletons, their great mobility and the fact that their physiological requirements seem to have changed little over the last two million years make Coleoptera ideal indicators of major climatic changes at a particular site (Coope 1975, 1977).

The sedimentary and fossil record of environmental change in lowland zones is often well-documented and has been the basis of the establishment of the continental Quaternary stratigraphic record. This record has been supplemented by geomorphological evidence in the form of moraines, outwash plains (sandar) and loess plains formed at either the maximum extent of the last ice sheet to cover the area or during retreat stages. Landform assemblages have also been widely used to determine the lines of movement of ice sheets and the mode of ice wastage. Drumlin fields, crag and tail, esker systems and meltwater channel systems all provide information about flowlines in the ice sheet and the manner of its dissipation.

The coastal zone

The present coastline is often a major source of information about Quaternary environments. Because vast volumes of water are withdrawn from the ocean basins and transferred to the ice sheets on the continents during a glacial period, world sea-level falls. At the same time, the sheer weight of the ice on the continents results in isostatic depression of those land masses. As deglaciation begins and the meltwaters return to the ocean, there is a lag in the rate of isostatic rebound of the continents and periods of relatively high sea-level occur during the early phases of deglaciation. The coastal zone illustrated in the model (Fig. 1.3c) is but one of the many possible sedimentary sequences which can occur as a result of a glacial cycle. The most significant aspect of the coastal zone stratigraphy is that it often contains rapid changes from fluvial to marine sediments as the position of the coastline changes in response to isostatic and eustatic fluctuations. In particular the molluscs and foraminifera included in coastal sediments reveal the changes from freshwater to marine conditions. In the sequence illustrated, the interglacial sands and gravels containing marine shells represent a relatively high stand of sea-level during the first interglacial. These marine deposits are overlain by barren fluvioglacial outwash deposits indicating a period of ice sheet expansion and a falling sea-level. The ice mass then overruns the coastal site and deposits till. During the early phases of deglaciation relative sea-level remains high and a deltaic sequence of fluvioglacial sediment accumulates in a marine environment. With the isostatic rebound continuing faster than the eustatic rise in sea-level consequent to the world-wide ice wastage, the deltaic sequence gives way to normal outwash deposition. As soon as the eustatic rise in sea-level outpaces the rate of isostatic land rebound a marine transgression occurs which produces the topmost marine sequence.

Fluctuations of relative sea-level can produce significant morphological changes in the coastal zone. Raised beaches, raised estuarine mud flats, abandoned cliff lines, sea caves, arches and stacks all bear testimony to sea-level fluctuations in areas which have been glaciated. Unless these landforms have fossil bearing sediments intimately associated with them it is often difficult to ascertain the relative or absolute age of the individual landforms. Many raised rock-platforms and abandoned clifflines are the product of more than one sea-level fluctuation. If former shorelines can be accurately dated then they do provide very important information about the extent of crustal warping as a result of glacial loading and they can be useful in determining the positions of ice margins during deglaciation.

The continental shelf zone

It has only been in the last twenty years, particularly as a result of mineral exploration in middle and high latitudes, that information has become available about Quaternary deposits and landforms on continental shelves. The expansion of ice sheets from the interior of continents across what is presently the coastline and on to the surrounding shelves is now well-established. A typical sedimentary sequence (Fig. 1.3) related to a glacial cycle would consist of an interglacial marine sequence with a warm water fauna being succeeded by a cold water fauna which is overlain by a glacial sequence consisting primarily of till. With the recession of the ice at the end of the glacial period the initial fauna would again be representative of cold water conditions but higher up the sequence it would be replaced by a warm water fauna. This very simple model would be greatly changed if, because of the shallowness of the shelf sea, the expanding ice mass did not traverse the shelf until after it became land as a result of the eustatic lowering of sea-level. The details of shelf stratigraphy are only slowly emerging as more bore-hole information becomes available. Similarly, details of glacial and fluvioglacial landforms in the continental shelf zone are very sparse.

The ocean basin zone

Whereas in continental areas the erosional activity of a glacial period may destroy the sediments of the preceding interglacial period the deep ocean basins are more likely to contain long sedimentary records. Over the last 30 years the examination of many deep-sea cores has proved that inter-core correlation is possible on the basis of their isotopic, palaeontological and magnetic characteristics (Emiliani & Shackleton 1974, Imbrie & Imbrie 1979). The sedimentary sequence in an ocean basin representing one glacial cycle may be only two to three metres thick. Detailed studies of the foraminifera present in such a sequence have been of enormous value in interpreting environmental changes in the Quaternary (see Bowen 1978, Chapter 3, and Imbrie & Imbrie 1979, Chapter 10). In a simple sedimentary sequence (Fig. 1.3d) the sediments representing the interglacial conditions would contain the foraminifer *Globorotalia menardii* but the fossil would be absent from the glacial sequence. If studies were made of the oxygen isotope ratios of the calcium carbonate skeletons of the foraminifera in the interglacial deposits it would be found that they contain higher concentrations of ^{16}O while the foraminifera deposited during the glacial period contain high concentrations of ^{18}O. These oxygen isotope variations between glacial and interglacial phases are believed to be the result of the changes in the isotopic composition of the oceans related to the preferential storage of the lighter oxygen isotope (^{16}O) in continental ice sheets during a glaciation. A plot of the changes in $^{16}O/^{18}O$ ratio against time represents a record of the expansion and contraction of the world's ice cover.

THE DATING OF QUATERNARY ENVIRONMENTAL CHANGES

All of the pioneer work in reconstructing Quaternary environments was based on sedimentary sequences exposed on the continents. The traditional stratigraphic method of investigation allowed dating by *relative* means: i.e. 'older than', or 'younger than'. The classical models of glacial and interglacial episodes developed in the European Alps (Penck & Bruckner 1909), in northern Europe (Keihlack 1926) and in central North America (Chamberlin 1884, 1895, Leverett, 1898,

1899) were all constructed on the basis of determining the relative ages of strata believed to represent either glacial or interglacial environments. There was much debate about the length of time occupied by each glacial and interglacial and it was not until the development of a range of techniques collectively known as geochronometric dating that the chronology of Quaternary environmental changes could begin to emerge. A review of the various dating techniques applicable to Quaternary studies is provided by Bowen (1978, Chapter 5) and only a brief statement about each of the techniques need be made here. The most significant dating techniques depend on natural rates of decay of radioactive isotopes and the first technique, developed by W. F. Libby, involved measuring the decay of unstable carbon-14 following the death of an organism. By measuring the radioactivity in a fossil sample (wood, charcoal, peat, shell, bones), comparing it with a modern standard and knowing its rate of decay (half-life), it is possible to calculate the time that has elapsed since death. Whenever Quaternary sediments contain suitable materials for radiocarbon analysis then a date can be obtained for the geological event associated with the sample. Dates are normally published with reference to a zero year which is taken as 1950 AD and are cited in terms of bp (before present i.e. 1950 AD). The radiocarbon technique is applicable over a time span of approximately the last 70,000 years but any contamination of the dated sample by modern carbon produces much larger errors on samples more than 30,000 years old, e.g. a contamination of one per cent in a sample of true age 40,000 will give an apparent age of 32,000 (Bowen 1978, p. 122). Although the radiocarbon technique has many problems it has proved extremely useful in dating environmental changes associated with the last glaciation and the Postglacial period.

Radiometric dating using the isotopes ^{230}Th, ^{234}U, ^{39}K, ^{40}Ar is also possible. The thorium technique can be used on organic carbonate and has a range of 0–200,000 years. The uranium technique can be used on corals and has a range of 50,000–100,000 years and the potassium-argon technique is applicable to volcanic rocks such as lavas and tuffs and some workers would restrict its use to materials older than 250,000 years (Shotton 1977).

The dating of Quaternary environmental changes has been a highly controversial topic during the last 20 years. The correlation of glacial and interglacial episodes over wide areas has been anxiously sought in an attempt to find explanations for the climatic changes responsible for them. The gradual erosion of the old, rather simplistic, four-fold subdivision of the Quaternary of Penck and Bruckner has taken place (Bowen 1978, Imbrie & Imbrie 1979) as more and more information has been obtained about the number, extent and chronology of climatic changes over the last two million years. Undoubtedly the contribution of the studies undertaken on cores of deep-sea sediments and the development of geochronometric dating techniques to provide dates for the major environmental changes revealed by the deep-sea sediments has revolutionised Quaternary studies. The contribution of oxygen isotope studies of foraminifera to produce a graph of the expansion and contraction of the world's ice sheets over the last 1·8 million years and the ability to date these events by means of measuring the palaeomagnetism of the sediments in the cores, which in turn has been dated by potassium-argon analysis of contemporary volcanic rocks, have together provided a chronology of environmental change throughout the Quaternary. There had long been a debate about the significance of the variations in the distance between the Earth and the Sun, changes in the axial tilt of the Earth and the precession of the Equinoxes in determining climatic changes on the Earth. In 1842, J. A. Adhemar suggested that the prime causes of ice ages might be variations in the way the Earth moves around the Sun. This suggestion was developed by a largely self taught Scottish scientist, James Croll. While employed as a janitor in the Andersonian College and Museum in Glasgow he started work on calculating the orbital eccentricity of the earth at various dates over the past three million years. He concluded that there must be something about a highly elongated orbit that causes ice ages every 100,000 years. Croll was able to demonstrate that the intensity of radiation received by the Earth during each season is strongly affected by changes in eccentricity. Croll was probably the first scientist to identify the effect of a 'positive feedback' because he argued that any astronomically induced change in solar radiation which produced an increase in winter snow accumulation would result in additional loss of heat by reflecting more heat back to space. As a result of his calculations, Croll suggested that individual ice ages last about 10,000 years and occurred first in one hemisphere and then in the other in response to the 22,000 year precessional rhythm. He identified eight ice ages in the northern hemisphere during the last glacial epoch which he believed lasted from 250,000 to 80,000 years ago (Croll 1875).

By the end of the nineteenth century Croll's theory of ice ages had largely been abandoned by the scientific community. However, Milutin Milankovitch set out to demonstrate mathematically that variations in orbital eccentricity and axial precession are large enough to cause ice sheets to expand and contract. In 1924 Milankovitch published a radiation curve (Fig. 1.4) for latitude 65°N over the last 600,000 years and indicated the four European ice ages on the graph (Koppen & Wagner 1924). The most valuable aspect of the Milankovitch theory was that it predicted how many ice age deposits geologists would find and the approximate age of the deposits—he predicted nine ice ages during the last 600,000 years.

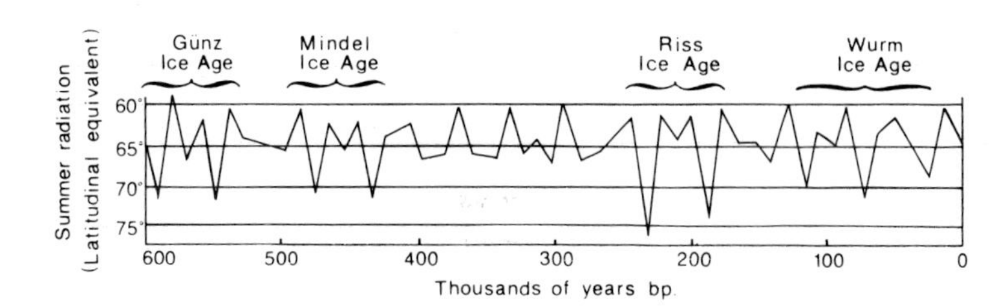

Fig. 1.4. Milankovitch radiation curve for latitude 65°N (Koppen & Wegener 1924). Variations in radiation intensity are expressed in terms of latitudinal equivalents, e.g. radiation received 590,000 years ago at 65°N is equivalent to that now received at 72°N.

The dating of magnetic polarity epochs by means of the potassium-argon method using terrestrial lavas (Cox *et al* 1963, Cox & Dalrymple 1964) was the fundamental breakthrough which allowed a timescale for the Quaternary period to be established and for the link between the Milankovitch theory of climatic changes and the geological evidence reflecting those changes to be established. In 1973 N. J. Shackleton and N. D. Opdyke published a graph (Fig. 1.5) of isotopic and magnetic measurements on a Pacific deep-sea core (V28-238). This graph, which showed that the Brunhes Epoch (the last 700,000 years) could be divided into nineteen isotopic stages reflecting changes in the light isotopes (^{16}O) in the ocean—not changes in water temperature—was an expression of the expansion and contraction of the world's ice sheets over this period, and the graph could be dated by radiocarbon dates at the top and by a magnetic reversal at the bottom. It was therefore possible to estimate the date of each isotopic stage by interpolation within the 700,000 year Brunhes Epoch. There had been eight major periods of expansion of the world's ice sheets over the 700,000 year period. It was argued that since sea-water was mixed rapidly by currents, any chemical change in one part of the ocean would be reflected everywhere within a thousand years, and therefore Shackleton and Opdyke's curve provided the first accurate chronology of global changes in Quaternary climate and could be used to test the Milankovitch theory. Subsequent work on cores from the Indian Ocean (Hays *et al* 1976) provided an isotopic curve for the last 500,000 years

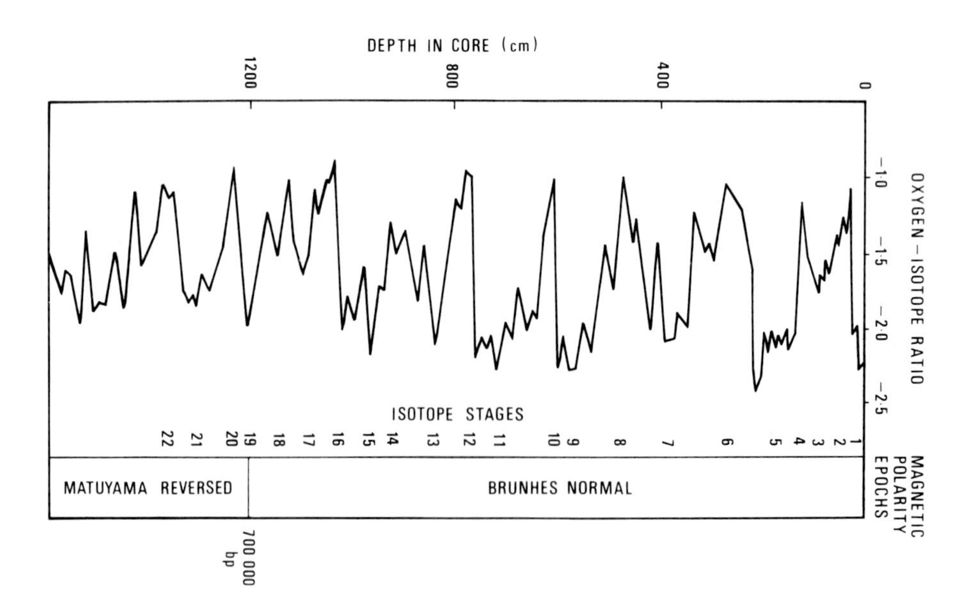

Fig. 1.5. Graph of oxygen–isotope ratios in Pacific deep-sea core V28-238 (Shackleton & Opdyke 1973).

and showed that climatic changes were synchronous in both hemispheres. Further work on the astronomical theory (Mesolella *et al* 1969, Berger 1977) revealed that climatic oscillations should occur as four distinct cycles: a 100,000-year cycle corresponding to variations in eccentricity; a 41,000-year cycle corresponding to variations in axial tilt; and 23,000 and 19,000-year cycles corresponding to variations in precession. Spectral analysis of the isotopic data from the ocean cores matched the predicted cycles within five per cent and it was concluded that the Quaternary ice ages had been triggered by changes in the Earth's eccentricity, precession and tilt.

The developments in Quaternary chronology and in understanding the causes of climatic change, described above, had two fundamental contributions to make to studies of environmental changes. Firstly, the number of glacial/interglacial cycles which occurred during the Quaternary was far larger than had been believed and secondly, that the importance of radiation variations was such that global changes in climate could be identified. It was also established that glacial periods lasted about 100,000 years and tended both to be initiated and terminated very rapidly. Interglacial (relatively warm) periods lasted only about 10,000 years.

THE HISTORY OF KNOWLEDGE RELATING TO ENVIRONMENTAL CHANGES IN SCOTLAND

At the beginning of the nineteenth century there was great interest in what were described as the 'superficial accumulations'—the masses of clay, sand, gravel and boulders—which occur throughout Scotland. Archibald Geikie states (1863, p. 2), "At a time when it was a common belief that the forces of nature once acted with a far higher intensity than they do now, the transport of the surface detritus, and the abrasion of the hills, were usually explained by reference to some vast currents or debacles of water, which, rushing across the country, tore up and transported huge quantities of sand and gravel, with large masses of solid rock." Although many papers were published (Bald 1809, Imrie 1812, Hall 1812, Macculloch 1812, Hibbert 1832) describing the drift deposits and explaining them in terms of vast bodies of water set in motion by violent earthquakes, as early as 1802 Playfair suggested that even the most violent floods could not transport huge blocks of stone and that (Playfair 1802, p. 349), ". . . for the moving of large masses of rock the most powerful agents, without doubt, which nature employs, are the glaciers."

A major contribution to the debate about the interpretation of the origin of the superficial deposits of Scotland was made by Louis Agassiz, the eminent Swiss naturalist who visited the British Isles during the autumn of 1840. Four years previously Agassiz had been converted to the glacial theory by Ignace Venetz and Jean de Charpentier, and after spending three summers studying glaciers in the Alps and examining the moraines, striations, polished surfaces, roches moutonnées and erratics produced by the Quaternary expansion of the Alpine glaciers he was ready to apply his concept of a former European ice sheet to the evidence available in other areas of Europe which did not contain any glaciers at present. He was quoted in *The Scotsman* of October 7, 1840 as stating, "After having obtained in Switzerland the most conclusive proof, that at a former period the glaciers were of much greater extent than at present, nay, that they had covered the whole country, and had transported the erratic blocks to the places where these are now found, it was my wish to examine a country where glaciers are no longer met with, but in which they might formerly have existed."

It was William Buckland, the Reader in Geology at Oxford who suggested that Agassiz should visit Britain in 1840. Buckland had visited Agassiz in Switzerland in 1838 and had accepted much of the glacial theory as it had been demonstrated by Agassiz in the field.

The details of the tour of the British Isles made by Agassiz in the autumn of 1840 have been presented by G. L. Davies (1968). Little is known about the first three weeks of Agassiz's stay in Britain but he arrived in Glasgow on or before September 21st because on that day he read a paper to Section D (Zoology and Botany) of the British Association for the Advancement of Science. On Tuesday September 22nd he read a paper entitled "On Glaciers and Boulders in Switzerland" to Section C (Geology and Physical Geography). The paper was delivered in French and Agassiz discussed the ability of glaciers to erode and transport rock debris and he claimed that ". . . at a certain epoch all the north of Europe, and also the north of Asia and America, were covered with a mass of ice." In the summary of the paper in the official report of the meeting (Rept. Brit. Ass., Glasgow 1840, pt II, p. 143), there was no specific mention of any glaciation of the British Isles but a report published in *The Athenaeum* contains the following passage: "Prof. Agassiz is also inclined to suppose that glaciers have been spread over Scotland . . . . If we understood him rightly, he means to follow up his valuable researches in the Highlands of Scotland during his stay in the country, and that he confidently expects to find evidence of such glaciers having existed." Based upon evidence that he collected at several building-sites in Glasgow and on his subsequent journey through the western Highlands with William Buckland, Agassiz soon became convinced of the former existence of glaciers in Scotland.

The chairman of the meeting of Section C of the British Association at which Agassiz presented his paper was James Smith of Jordanhill. Smith is often referred to as the 'yachtsman geologist' because it was during his sailing activities in the Firth of Clyde that he discovered the arctic fauna of the raised estuarine clays (Clyde Beds) which outcrop around the shores of the Firth. In 1839 Smith had published a paper based on his analysis of the arctic species to be found in the Clyde Beds and he had concluded that Scotland had in the recent geological past experienced a much more severe climate than at present.

After the British Association meeting in Glasgow, Agassiz and Buckland travelled north via Inverary, Pass of Brander, Loch Etive, Loch Creran, and Ballachulish to Fort William. From Fort William they visited Glen Spean, Glen Trieg and Glen Roy. In Glen Roy they saw the famous 'parallel roads' which, in the previous year (1839) Charles Darwin had suggested were the product of marine activity. Agassiz wrote the following letter at Fort Augustus on October 3rd and it was published in *The Scotsman* on October 7th: ". . . at the foot of Ben Nevis, and in the principal valleys, I discovered the most distinct moraines and polished rocky surfaces, just as in the valleys of the Swiss Alps, in the region of existing glaciers, so that the existence of glaciers in Scotland at early periods can no longer be doubted. The parallel roads of Glen Roy are intimately connected with this former occurrence of glaciers, and have been caused by a glacier from Ben Nevis. The phenomenon must have been precisely analagous to the glacier-lakes of the Tyrol, and to the event that took place in the valley of Bagne."

It is clear, therefore, that by October 1840 both Agassiz and Buckland were convinced of the former existence of glaciers in Scotland and Buckland wrote to Agassiz on October 15th in Ireland that Lyell had also accepted this view. On November 4th Agassiz read a paper entitled "Glaciers and the Evidence of their having once existed in Scotland, Ireland and England" to the Geological Society of London, with Buckland, the President, in the chair. It was a rather vague paper but was followed by papers presented by Buckland and Lyell which were closely argued deductions from field evidence collected in the north of England and Scotland during the preceding summer, in support of the Glacial Theory. Despite the great interest shown in Agassiz's visit to Britain and the papers he read in Glasgow and London it was some twenty-five years before the idea that an ice sheet and valley glaciers had occurred in Scotland in the recent geological past replaced the concept of recent marine submergence and iceberg activity as a reasonable explanation for the drift deposits.

There was a period (circa 1840–1865) when numerous papers were published in which the authors mentioned the possibility of a former ice sheet cover of Scotland while retaining the concept of marine submergence and iceberg activity—e.g. T. F. Jamieson's 1860 paper entitled "On the Drift and Rolled Gravel of the North of Scotland." However, by 1862 the same author in a paper entitled "On the Ice-Worn Rocks of Scotland" stated (p. 166) ". . . the grinding down of the rocks has, in Scotland at least, been caused by the long continued movement of land ice and glaciers." Jamieson argued strongly against the significance of sea ice and icebergs and supported Agassiz's suggestion that the parallel roads of Glen Roy were the shorelines of an ice-dammed lake.

Perhaps the most significant publication relating to Quaternary environments in Scotland after the introduction of the Glacial Theory to Britain in 1840 was that of Archibald Geikie (1863) *On the Phenomena of the Glacial Drift of Scotland.* He admits in the preface of that publication (p. iii) that, "It was not until the summer of 1861 that I finally abandoned the attempt to explain the origin of the rock dressings and the boulder clay by the action of icebergs." Geikie gives an excellent summary of the papers published on the Scottish 'drifts' during the period up to 1863 and points out that British geologists were reluctant to admit that ice sheets could ever have covered the lowlands. Based on detailed fieldwork Geikie was able to describe many sections and landforms which provided a basis for a chronological description of the

Scottish environment;

"I. The Old or General Glaciation of the country
II. The marine submergence and re-elevation
III. The final disappearance of the ice."

He clearly distinguished between the effects of the ice sheet (I) which was followed by a marine transgression to produce the marine clays with their arctic fauna and a final phase of valley glaciers (III). Geikie (1863, p. 94) states, "The occurrence of stratified deposits in the Scottish till further indicates that though the intervals of quiet and vegetation may have lasted for many years, they were in the end brought to a close by the return of severe winters, when the ice once more crept down the valleys which it had abandoned."

In 1865 T. F. Jamieson published an excellent paper entitled "The History of the Last Geological Changes in Scotland." The paper included a map (Fig. 1.6)—probably the first Quaternary map of Scotland—which shows the chief ice sheds, lines of ice movement and the location of fossiliferous glacial marine beds and the carse clays. Jamieson certainly believed in the importance of land ice but retained the concept of a great marine submergence and obviously felt required to state (p. 165), "At the same time I do not mean to deny that there has been some scratching by means of floating ice". He also suggested that the great marine submergence may have been caused by the weight of the ice sheet on the land. Jamieson's conclusions about the environmental changes recorded in the superficial deposits in the area between the Firth of Forth and the Moray Firth are really the foundations upon which all subsequent work on Quaternary environmental changes in Scotland are based and are worthy of quotation in full:

"1st. After the deposition of the Crag-gravel, and after the Mammoth had lived in Scotland, the country was covered with a great depth of snow and ice, which must have extinguished the pre-existing flora and fauna. This ice moved outwards in broad streams from the great watersheds of the country, carrying with it much stony debris and multitudes of boulders, which it left in irregular sheets, constituting the old boulder clay or "till" of some authors; the ice also scratched and furrowed the rocks, destroyed the pre-existing alluvium, and exercised a considerable amount of abrasion on the surface of the country.

2nd. After this state of things had continued for a time, a depression of the land took place to the extent of some hundreds of feet, so that all the lower grounds were below the sea-level, but as to the full extent of the depression we are still ignorant. During this submergence the brick-clays containing Arctic shells were deposited, boulders were drifted here and there by floating ice, and it seems probable that the ice still covered much of the land, and even protruded in to the sea along the main valleys in the form of large glacier-streams; so that the condition of the country would have been like the present state of Spitzbergen.

3rd. The country emerged from the water, but ice still lay on much of the land, and perhaps reoccupied some of the tracts over which the sea had spread, deranging by its intrusive action the marine beds of the preceding period.

4th. The glaciers at length began their final retreat, leaving behind them heaps of rough debris and mounds of gravel, more especially at those points where they halted for a time. Large quantities of rolled gravel were also strewed along the valleys by the water issuing from beneath the ice, and by the floods occasioned by rapid thaws, the absence of vegetation on much of the surface probably contributing to the effect.

5th. By this time the land attained a higher level than it has at present, so that the area of Britain was much larger than it is now, and, instead of presenting the appearance of a group of islands, formed a mass of connected land united to the Continent of Europe, the flora and fauna of which now spread into it. Woods of Birch, Hazel, Alder, and other trees covered the surface, and the Great Irish Elk, the Red Deer, the Great Wild Bull, the Wolf, the Bear, the Beaver, and probably the Reindeer, were amongst its

Fig. 1.6.

Sketch-map illustrating the glacial phenomena of Scotland (Jamieson 1865).

Explanation—"The arrows show the directions of the glacial markings; the thick black lines the chief ice sheds, at lines whence the land-ice flowed during the period of the Boulder-earth. The ruled parallel lines show the districts where the Brick-clays and fossiliferous glacial-marine beds seem chiefly to occur. The dotted lines illustrate the distribution of the valley-gravel; the black patches mark the beds of old estuarine mud, or Carse-lands; and the sites of submerged forests are indicated by a cross (+)".

inhabitants. In the valleys the rivers were gradually cutting their way through the masses of glacial debris to lower levels, and in doing so spread out much gravel and alluvial soil along their banks. This period is represented by the submarine forest and bed of peat underlying the Carses of the Tay and Forth.

6th. A depression now took place, cutting off the land-connexion with the Continent, isolating Ireland and the various islands, and thus stopping the land-migration from Europe. In the valley of the Tay and Forth this old coast-line was 25 or 30 feet above the present, but on the coast of Aberdeenshire not beyond 8 or 10. The old estuarine beds, or Carses, of the Forth, Tay, and other rivers were formed, together with corresponding shingle-beaches and caves along the coast. Man having by this time got into the country, evidence of his presence appears in the shape of canoes and primitive weapons of stone and horn buried in deposits of the period.

7th. A movement of elevation (whether gradual or sudden is uncertain) at length took place, so that the land attained its present position, thereby laying dry the Carse districts and old coast-line. Since this occurred much peat has been formed, and a great amount of blown sand has been heaped up on certain parts of the coast. In some districts the natives continued for a time to use tools and weapons of flint and stone, and left shell-mounds in the neighbourhood of the estuaries. Some of the wild animals were gradually extirpated, such as the Great Wild Bull, the Bear, the Beaver, the Wolf, and the Capercailzie,—the Great Elk and the Reindeer having probably disappeared at an earlier period. Since the dawn of Scottish history, and the occupation of the lowlands by the Saxon race, no noticeable change of level has been observed."

The conclusions reached by Geikie (1863) and Jamieson (1862, 1865) mark a new era in the interpretation of Quaternary environmental changes in Scotland—they demonstrated that the last ice sheet had completely covered Scotland and that its wastage was associated with a relatively high stand of sea-level. After a period of ice retreat, valley glaciers expanded for a short period of time and relative sea-level fell to allow peat to form at the same time that vegetation was colonising the newly deglaciated landscape. The peat of the coastal lowlands and estuaries was then submerged by the main Postglacial marine transgression and the carse clays were deposited. The present sea-level was subsequently attained and peat developed on the abandoned carse clays. The relative chronology and significance of each of these events have been little changed by a further 120 years of research. Some of the fundamental observations and conclusions reached by Geikie and Jamieson were in fact ignored by subsequent workers, e.g. Jamieson clearly believed that the raised beaches had been affected by differential uplift of the land and that a beach of a given age did not attain the same altitude above present sea-level at all locations.

For some thirty years after 1865 the brothers Archibald and James Geikie dominated the field of Quaternary studies in Scotland. Sir Archibald Geikie was Professor of Geology and Mineralogy at the University of Edinburgh and subsequently became Director-General of the Geological Survey of the United Kingdom. The two editions of his book entitled *The Scenery of Scotland* (1865, 1901) include chapters on ice sheets and glaciers and on the changes in landforms, vegetation and sea-level which have occurred in the Postglacial period. The three editions (1874, 1877, 1899) of James Geikie's book, *The Great Ice Age and its Relation to the Antiquity of Man* are clearly aimed at interpreting the climatic changes in the Glacial Epoch. The books are wide ranging and include chapters not only on glacial and postglacial phenomena of Scotland but also of Greenland, Scandinavia, England, Switzerland and North America. Consideration is given to glacier motion, causes of climatic change and the archaeological evidence associated with the drift deposits. The twenty year gap between the first and third editions reveals a steady progress in the discovery of more information about the effects of Quaternary environmental changes on the Scottish landscape. In the third edition, for example, there is a detached coloured map showing the principal lines of movement of the last ice sheet.

For the first half of this century the officers of the Geological Survey and individual scientists continued to fill in the details of the Quaternary story using the framework established by the Geikie brothers and Jamieson. Important individual contributions were made by Bremner (1932, 1934a, b, 1939), Charlesworth (1926a, b), and J. B. Simpson (1933). Not until 1954 was another review of the knowledge relating to Quaternary environmental changes in Scotland attempted by A. D. Lacaille (Archaeologist to the Wellcome Historical Medical Museum, London). This excellent book draws widely on the botanical, geological and archaeological evidence not only from Scotland but from throughout Europe to establish the sequence of environmental changes. Two aspects of the book remain as major contributions. It was the first book on the Scottish Quaternary to make extensive use of palynology and varve chronology. Lacaille placed the maximum of the last Scottish ice sheet at 20,000 bp and the Valley or Moraine glaciation (Loch Lomond Advance) at between 9,800 and 10,300 bp, and the Postglacial marine transgression at about 7,000 bp. He was able to make use of pollen diagrams produced by Erdtman (1928) and Godwin (1940) to reconstruct the general trends in the Postglacial vegetation changes which had occurred in Scotland, but he did not have at his disposal any radiocarbon dates. It was not until the early sixties that radiocarbon dates became available for Scottish sites (Sissons 1967a, p. 144).

The last twenty years have seen considerable advances in our understanding of the late Quaternary environments in Scotland. Sissons (1967a and 1976a) has produced overviews of the major developments in glacial geomorphology and relative sea-level changes. This book is concerned not only with the geomorphological and marine sedimentary record but with the broad environmental changes which can be identified as a result of detailed investigations not only in those fields but in all aspects of the geomorphological, stratigraphical, palaeontological and archaeological record.

NORTH-EAST ATLANTIC QUATERNARY ENVIRONMENTAL CHANGES

It is necessary, before discussing the environmental changes which occurred in Scotland over the last 30,000 years to place these changes in the wider context of the environmental changes believed to have occurred during the Quaternary Period in the northern hemisphere, and specially in the north Atlantic. In concluding his recent review of the stratigraphic framework of the Quaternary, Bowen (1978, p. 193), writes, "There is no doubt whatsoever that Quaternary systematics have been subject to a revolution comparable to that of plate tectonic theory on geology as a whole. Instead of the Pleistocene comprising four, five or at most six major glacials, it is fact that eight glacials and eight interglacials occurred during Brunhes time [last 700,000 years] alone, while altogether some seventeen glacials represent the entire Quaternary [circa 1·6 million years]." This has meant that most existing classification systems either require redefinition or should be abandoned. Since the number of glacial and interglacials has been largely determined from the oceanic record and continental sequences contain actual hiatuses, a major problem exists in correlating the oceanic and continental records and interpreting both sets of evidence in terms of environmental changes. Shackleton & Opdyke (1973) have proposed that Core V28-238 from the Solomon Plateau in the Pacific be used as the standard for the last 900,000 years. Bowen (1978, p. 198) has used that core as the basic stratigraphic framework for correlating events in various parts of Europe. Bowen points out that correlation with the North American continental stratigraphy is as yet not possible.

Of greater relevance to the study of environmental changes in Scotland during the Middle and Upper Quaternary are the results of detailed studies of the deep-sea biostratigraphy of the North Atlantic. The pioneer work of Schott (1935) showed that high percentages of *Globorotalia menardii* in oceanic deposits signified climatic amelioration. Lidz (1966) showed that climatic curves based on ratios of a group of species rather than single species agreed much better with oxygen isotope curves, and this approach was further developed by Imbrie & Kipp (1971) who established parameters for existing foraminiferal assemblages in modern sediments (summer and winter temperature and salinity values) which by means of transfer functions could be applied to fossil assemblages related to earlier environments. Ruddiman & McIntyre (1976) defined ecological water masses in terms of the various fossil assemblages and they demonstrated that eleven major advances of polar water occurred in the past 600,000 years in the North Atlantic. Ice rafted debris was also recorded in the bore-holes in association with the 'polar' fossil assemblages. These environmental changes have been dated using radiocarbon assay, interpolation from Uranium series dates and identification of a volcanic ash layer which erupted about 9300 bp.

One foraminiferal species—*Globigerinoides pachyderma*—characteristic of cold water conditions has proved particularly sensitive (Ruddiman, Sancetta & McIntyre

1977). A curve showing the percentage of this species present (Fig. 1.7A) in relation to the total planktonic foraminifera larger than 149 mm can be regarded as a temperature curve for the site of core K708-7 (53°56′N, 24°05′W). This approach has been extended to cover the past 600,000 years (Ruddiman & McIntyre 1976) and they tentatively conclude that there were eight full climatic cycles in the Brunhes epoch (B–I) and the beginning of a ninth (A)—Fig. 1.7B. There were also short but severe climatic pulses lasting less than 20,000 years and if these are included there were eleven major polar water advances since 600,000 bp. Because there are significant amounts of coarse, ice-rafted terrigenous debris present at each of these maxima the authors infer that continental ice sheets may have reached their full size in each of these cold periods and, if the record is extended back to 1·2 million years (Kent *et al* 1971), there may have been 20 or more significant ice sheet advances. The deep-sea sediments and their included fossils to the west of the British Isles suggest numerous and rapid climatic changes during the last 1·6 million years and point to the likelihood of at least eleven and probably sixteen major cold periods which affected Scotland.

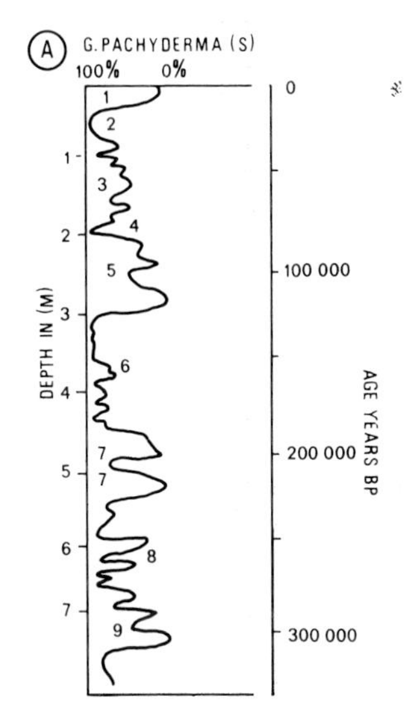

Fig. 1.7A. Polar fauna as a percent of total planktonic foraminifera larger than 149 mm—core K708–7. (Ruddiman *et al.* 1977.)

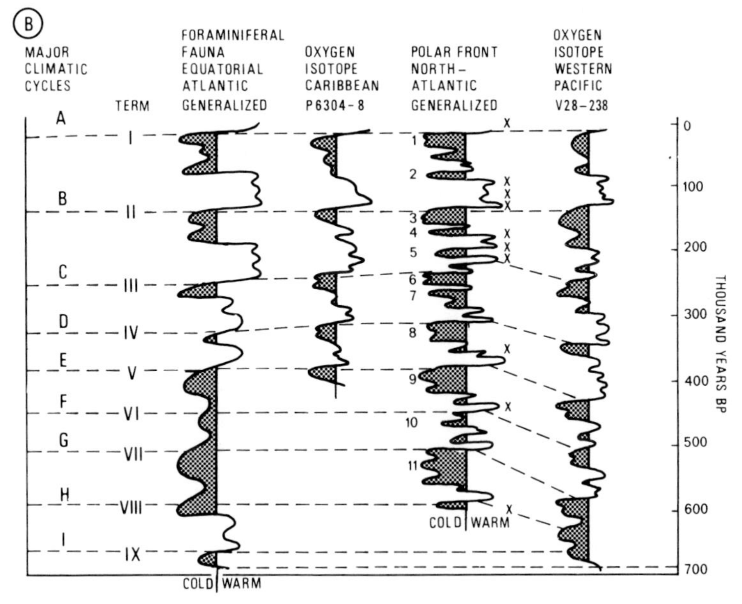

Fig. 1.7B. Comparison of cyclic trends in deep-sea cores (Ruddiman & McIntyre 1976).

Much more information is available about changes in oceanic conditions in the North Atlantic during the last 130,000 years (McIntyre *et al* 1972; McIntyre & Ruddiman 1972, Sancetta, Imbrie & Kipp 1973, Ruddiman & McIntyre 1973; McIntyre *et al* 1976; Kellogg 1976; Van Donk 1976; Ruddiman, Sancetta & McIntyre 1977; Ruddiman & McIntyre 1981). The locations of the cores from which the evidence for the changes in oceanic conditions have been derived are shown on Fig. 1.8. Core V23–82 (Sancetta *et al* 1973) revealed the following temperature conditions (Fig. 1.9)—a rapid warming began about 127,000 bp and continued until 73,000 bp but there were significant cooling episodes at 109,000 and 92,000 bp. The period 73,000–11,000 bp was generally cold with short warm intervals at 31,000; 45,000; and 59,000 bp. The rapid warming at about 11,000 bp was followed by a short sharp cold episode about 10,000 bp before the rapid rise in temperature associated with the beginning of the present climatic regime about 10,000 bp.

More detailed information has recently become available (Ruddiman *et al* 1977, Ruddiman *et al* 1978, Ruddiman & McIntyre 1981a, b) regarding the rate of change in oceanic conditions in the North Atlantic and the relationship between those oceanic conditions and the build up and wastage of glacier ice on the surrounding land masses. It has been established that sea surface temperatures in the North-east Atlantic have changed by as much as 7–11°C in response to large scale palaeoclimatic changes. Rates of cooling or warming usually average 1–5°C/1,000 years for a complete climatic shift, but during the passage of the polar front over a particular site, rates of change can be as high as 13°C/1,000 years. It would appear therefore that the position of the North Atlantic polar front (separating polar from subpolar water) is crucial to the environmental conditions in Scotland (Fig. 1.10). At the time of the last Interglacial (circa 127,000 bp) the polar front stood close to the south-east coast of Greenland. At 65,000 bp it was 500 km north-west of Scotland and at the maximum of the last glaciation it was south of the Bay of Biscay.

A major question relating to the expansion of northern hemisphere ice sheets after the last Interglacial has been the rate and chronology of that expansion. Ruddiman *et al* (1978) have demonstrated that by 75,000 bp (Fig. 1.11) at least *75* per cent of the total ice expansion of the last glacial episode had been achieved and yet there was marked oceanic warmth in the North Atlantic and sea-surface temperatures off the west coast of Scotland were similar to those of today. In fact, it was this warm ocean which provided a source of moisture for ice sheet expansion on the adjacent continents. It was not until after 75,000 bp that winter sea-surface temperatures began to fall below 1°C at the site of core V23–82 (Fig. 1.9) and it might be suggested that apart from five short (1,000 years) sharp cold episodes the environment of the north-eastern Atlantic was more characteristic of interglacial conditions rather than glacial conditions from 127,000 to 72,000 bp. The period between 72,000 to 30,000 was generally cold except for three periods each of less than 3,000 years when temperatures rose by from 3–6°C at about 58,000 bp, 48,000 bp and 32,000 bp. The greatest warming (6°C) occurred about 48,000 bp and can be correlated with the Upton Warren Interstadial in England (Fig. 1.9). Whether or not glacial ice occurred in the British Isles between 72,000 bp and

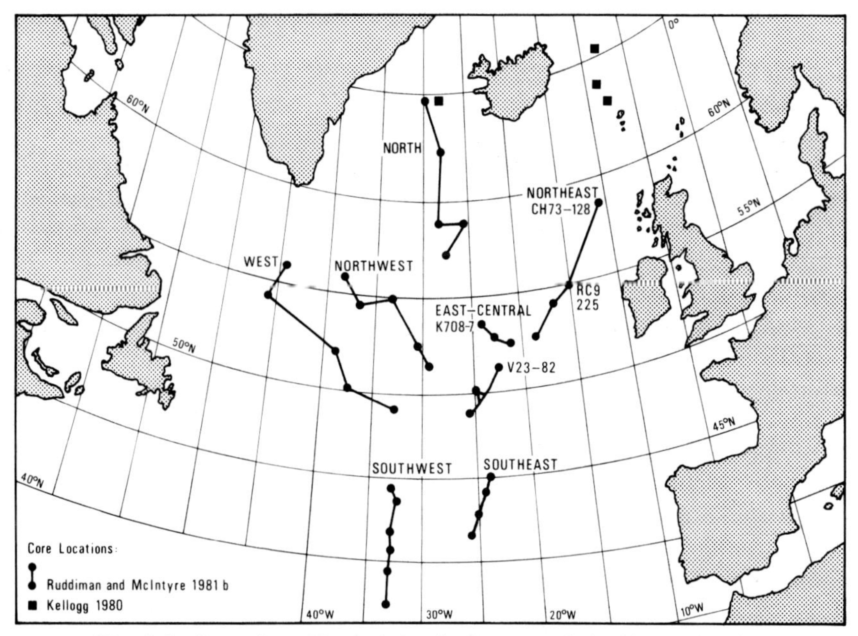

Fig. 1.8. Core sites, North Atlantic Ocean and the Norwegian Sea.

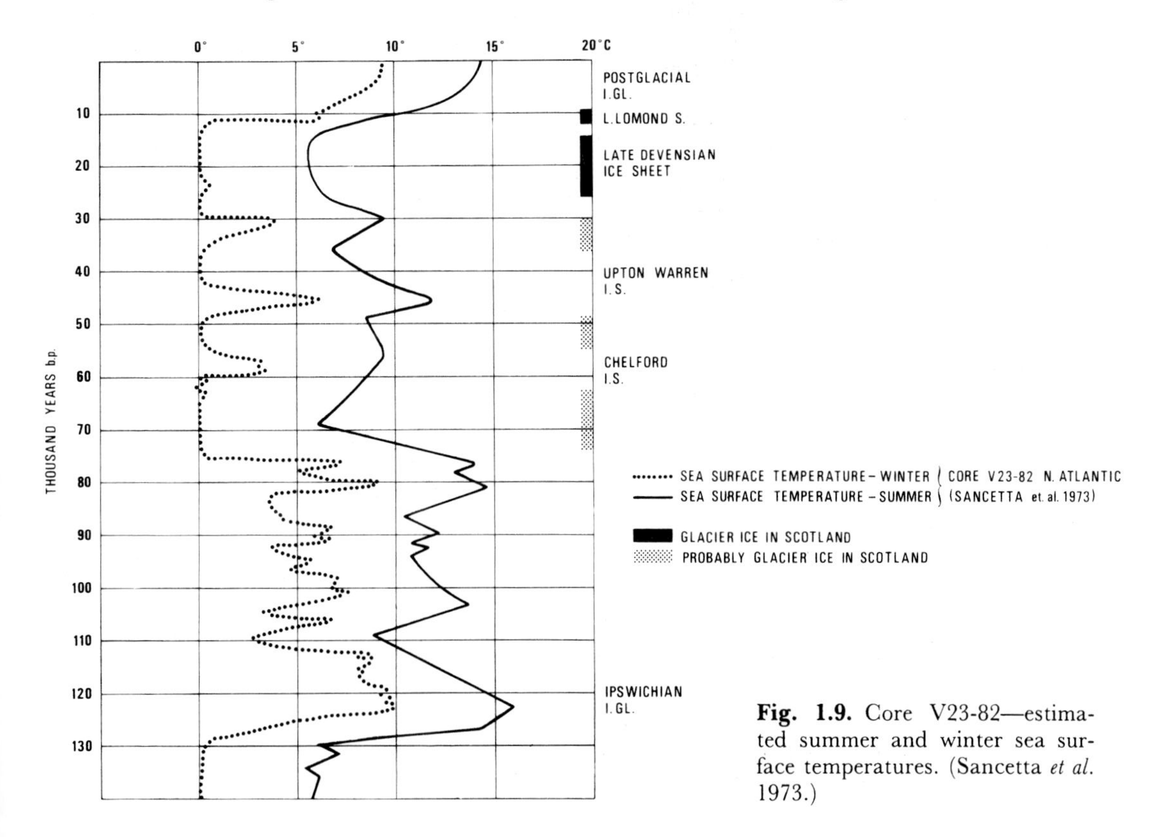

Fig. 1.9. Core V23-82—estimated summer and winter sea surface temperatures. (Sancetta *et al.* 1973.)

30,000 bp is not known, but since winter temperatures in the north-east Atlantic only fell below 3°C for only very short periods it seems unlikely that extensive glacierisation did occur.

Although the palaeotemperature estimates derived from the study of ocean cores in the north-east Atlantic are extremely useful in suggesting environmental trends in Scotland, it is not possible to put too much reliance on the chronology of the environmental changes. Sancetta *et al* (1973, p. 113) state, "All of the dates for climatic events . . . represent linear interpolations . . . rounded to the nearest thousand years. They are to be taken as best estimates, with maximum uncertainties varying from level to level." The chronology for core V23–82 (Fig. 1.7) has uncertainty estimates at 127,000 bp of ±6,000, at 73,000 of ±5,000, at 19,600 of ±2000 and 10,200 ± 1,000.

The environmental changes in the northern Atlantic during the last deglaciation (18,000–10,000) are well documented (Ruddiman & McIntyre 1981a, b). The authors have used the interpretation of faunal species composition and a salinity/productivity interpretation of faunal abundance changes to identify ecological watermasses during five major phases of deglaciation::

Phase I (16,000 bp). The ice volume maximum occurred around 18,000 bp—the date

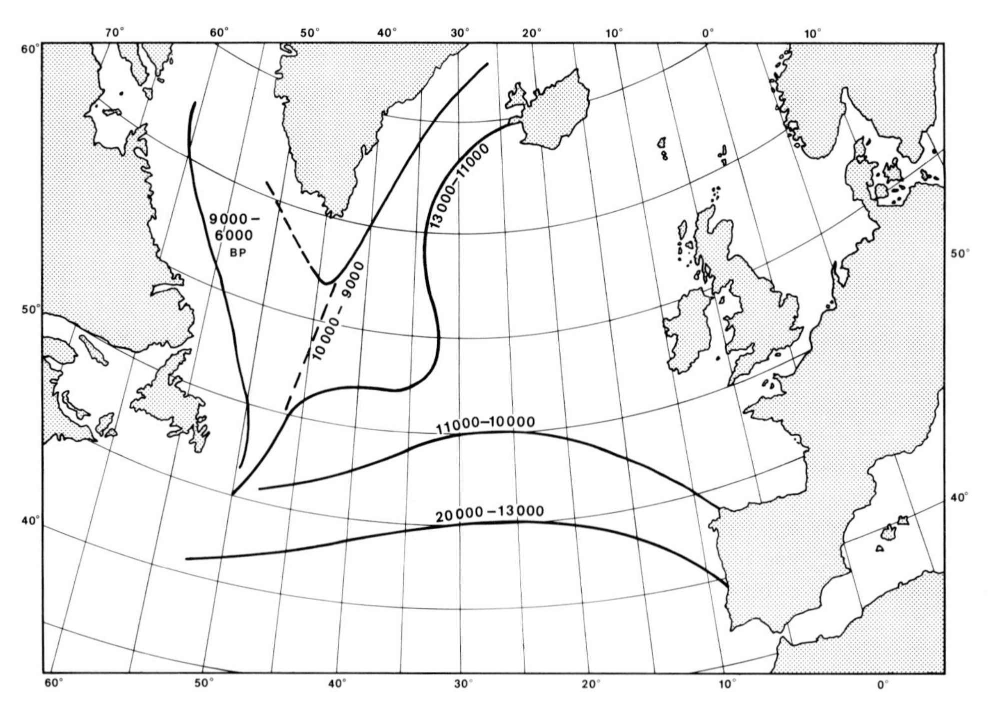

Fig. 1.10. Retreat positions of the North Atlantic Polar front from the glacial maximum position 20,000 years ago to the modern interglacial location after 6,000 years bp (Ruddiman & McIntyre 1981).

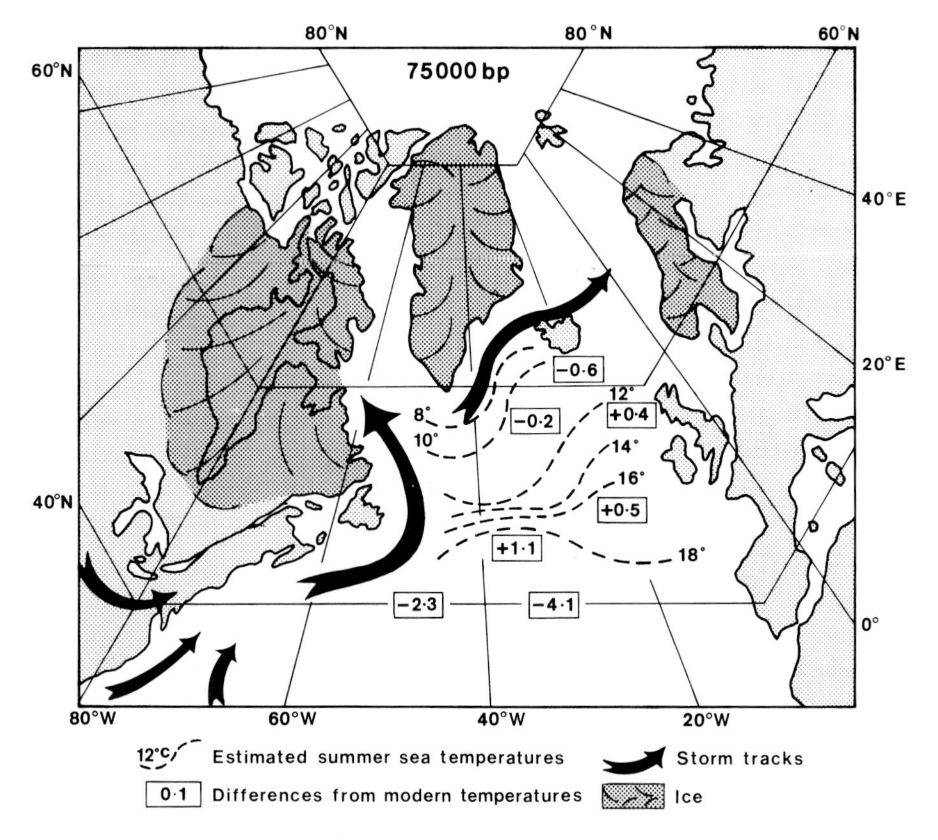

Fig. 1.11. Inferred main storm tracks and estimated summer sea surface temperatures and glacier ice cover at 75,000 bp (Ruddiman *et al.* 1980).

of the ^{18}O maximum values in the benthic and planktonic foraminifera. There is no evidence of any regions north of 45°N in which *Globigerina pachyderma* (polar species) does not reach greater than 90 per cent dominance. This was a phase of severe cold.

Phase II (*16,000–13,000 bp*). This was the period of the maximum influx of melt-water and icebergs from the continental ice sheets. This is expressed by high percentages of ice-rafted sand-sized non-carbonate sediments in several North Atlantic cores. The North Atlantic was cold and barren for some 4,000 years after the beginning of the retreat of the ice sheets on the continents about 17,000 bp.

Phase III (*13,000–9,000 bp*). The beginning of this phase is estimated by interpolation and extrapolation of sedimentation rates and is not precise. The initial warming is estimated to have begun at 13,000 bp and it is not possible to estimate winter sea ice limits during the period after 13,000 bp. The east-central subpolar Atlantic became markedly warmer between 13,000 and 11,000 bp with the polar front retreating far to the west and north (Fig. 1.10). There was a sharp reversal in the deglacial warming between 11,000 and 10,000 bp with a cooling in the sea-surface temperatures of approximately 10°C. This cooling is equivalent in age to the European Younger Dryas (Loch Lomond Advance, or Lateglacial Stadial in Scotland). From 10,000 to 9,000 bp much of the North Atlantic warmed to nearly full Postglacial temperatures and the polar front retreated far to the north and west (Fig. 1.10).

It would appear that at least 50 per cent of the total last glaciation ice volume of the northern hemisphere had been dissipated by 13,000 bp, and that there was a marked slowing down in the pace of northern hemisphere deglaciation after that date. It is probable, however, that this 50 per cent hemispherical deglaciation was sufficient to allow Scotland to become ice free by about 13,000 bp. The entire North Atlantic ocean surface south to at least 50°N was flooded by meltwater and icebergs in summer and probably covered by sea ice in winter between 16,000 and 13,000 bp. The winter cyclonic storm tracks would therefore be south of the 50°N sea ice limit and the Fennoscandinavian and Scottish ice sheets were denied winter moisture. The wastage of these ice sheets would be the product of rising temperatures accompanied by moisture starvation—the deglaciation took place in a semi-arid environment. Van Geel & Kolstrup (1978) state that the Lateglacial climate of Europe was one of very dry, cold winters and dry but moderately warm summers. This interpretation reconciles the occurrence of warmth-indicating plants despite the lack of tree cover. During the period 13,000–9,000 bp most of the subpolar North Atlantic Ocean was less influenced by meltwater, less covered by winter sea ice and considerably warmer, except for the brief return to almost full glacial cold conditions between 11,000 and 10,000 bp. The oceanic data suggest that this millennium was the only time in the entire deglacial interval from 18,000 to 6,000 bp when very cold oceanic temperatures coincided with a moderate winter moisture flux from an ocean which, south of Iceland, was relatively free of sea ice.

It appears, therefore, that the oceanic evidence points to a rapid disintegration of the last Scottish ice sheet between 17,000 and 13,000 as a result of 'marine downdraw'. That is, rapid calving of ice into a rising relative sea-level, accompanied by moisture starvation in the interior of the ice sheet. The climate of the deglaciation period in Scotland was much drier than at present. The episode of climatic amelioration was terminated by the return of full glacial conditions coincident upon the southward migration of the polar front about 11,000 bp (Fig. 1.10). Ruddiman & McIntyre (1981b) have suggested that this southward migration of the North Atlantic polar front was caused by the break up and outflow of large ice shelves from the Arctic Ocean (Mercer 1969) which was possibly enhanced by external forcing (radiation levels) acting at a frequency close to 2,500 years (Denton & Karlen 1973).

The above review of the literature relating to palaeotemperature estimates for the North Atlantic Ocean over the last 130,000 years reveals two important conclusions. Firstly, there have been rapid and large changes in sea-surface temperatures over that period. These changes are in turn related to migrations in the polar front which have also occurred at great speed. The character of the oceanic waters which wash on to the west coast of Scotland have changed rapidly and frequently over the last 130,000 years. The amplitude of sea-surface temperature changes has been of the order of 10°C and such changes which have taken place have been reflected in the environmental conditions prevailing in Scotland. Both the magnitude and frequency of such changes have been much larger than previously believed.

The work of Ruddiman and McIntyre has also revealed that changes in oceanic temperatures are diachronous. Whereas the glacial/interglacial cycles revealed by oxygen-isotope analysis suggest that climatic changes on a global scale can be regarded as synchronous events, the fact that a migrating polar front can bring

TABLE 1.1

Environmental changes in the north-east North Atlantic and some British Correlations (Sancetta *et al* 1973, Bowen 1978).

Years bp	Sea surface temperatures (°C) V23–82		British correlations	
	JAN	JULY		
0	10·5	15·0	Flandrian I.G.	Postglacial
10,000	5·4	12·9	D	Lateglacial Stadial (Loch Lomond Stadial)
11,000	6·8	13·2	E	
11,500			V	Lateglacial Interstadial
	0·4	6·4	E	
21,800			N	Late Devensian (last Scottish) Ice Sheet
31,000	3·9	10·5	S	
36,000	0·2	6·8		
48,000	6·7	12·1	I	Upton Warren Interstadial
51,000	0·3	7·6		
59,000	3·7	9·6	A	Chelford Interstadial
73,000	0·5	6·3	N	
81,000	9·3	15·2		
109,000	2·5	9·0		
124,000	9·7	16·0	Ipswichian I.G.	
127,000	0·6	7·7		
250,000			Wolstonian GL.	
500,000			Hoxnian I.G.	Middle Pleistocene
			Anglian GL.	
700,000			Cromerian I.G.	
				Early Pleistocene
1·6 million				

major climatic changes to the shores of north-west Europe at different times depending upon the location of those shores in relationship to the axis of migration of the front, would suggest that climatic changes (and therefore other environmental changes) can be diachronous. The classic example of such diachroneity would be the chronology of the rapid warming at the end of the Lateglacial Stadial (Loch Lomond Stadial, Younger Dryas Stadial) within the British Isles. The north-westward migration of the polar front brought rapid warming and the beginning of the Postglacial period to south-west Ireland at least 500 years earlier than in north-west Scotland, (based on rates of migration of the polar front in the North Atlantic.)

The synchroneity of global climatic changes over periods of 100,000 years or more and the diachroneity of climatic changes over periods of 1,000–10,000 years are not incompatible. Unfortunately the problem of establishing the rates of climatic change over time periods of less than 10,000 years is made difficult by the possible inaccuracy of radiocarbon and other geochronometric techniques. Even the detailed studies of deep sea cores still have a question mark against their chronologies which are based on extrapolated sedimentation rates and selected radiocarbon dates from calcium carbonate rich levels. Both the oceanic and continental sedimentary records still requires more precise dating.

It is now possible to place British Quaternary environmental changes within the framework of the chronology established by the study of Quaternary sediments in the North Atlantic Ocean (Table 1.1). The oceanic records suggest many rapid environmental changes during the Early and Middle Pleistocene which have yet to be recognised in the continental record. However, the oceanic data for the last 130,000 years (Upper Pleistocene) is more easily correlated with the British sedimentary sequence. The remainder of the book is primarily concerned with an analysis of the record of environmental changes which have taken place in Scotland during the last 30,000 years.

2

SCOTLAND BEFORE THE LAST GLACIATION

Eighteen thousand years ago Scotland was completely covered by glacier ice. The development and wastage of this ice sheet will be discussed in chapter three but the implications of its existence on our understanding of environmental conditions in Scotland before the last glaciation are very significant. The interpretation of palaeoenvironments is heavily dependent upon the occurrence of fossil-bearing sediments associated with those environments. An episode of intense glaciation tends to remove most, if not all, of the terrestrial sedimentary record of earlier glacial or interglacial episodes. It is, therefore, not surprising that there is very little evidence available in Scotland relating to environmental conditions which existed throughout 95 per cent of the Quaternary period.

The principal landforms of Scotland developed over a time-scale of tens of millions, or in some areas, hundreds of millions of years. Undoubtedly the volcanic and tectonic activity of the Tertiary period had a profound effect on landform development and it is highly likely that the main valley systems were developed in the Late Tertiary period, that is after 50 millions bp (Sissons 1976, p. 28). Lack of sedimentary evidence makes the interpretation of landform development during the Tertiary period in Scotland extremely difficult. Some workers (e.g. Linton 1951a, and Godard 1965) have suggested that Tertiary climates in Scotland were considerably warmer than the present climate. It must be admitted that the lack of evidence of environmental conditions and of geomorphological evolution during the Tertiary is the result of the intense and frequent climatic changes which took place during the Quaternary period.

It is now established that there have been at least 16 major cold periods during the last 1·6 million years and eight glacials and interglacials during the last 700,000 years (Bowen 1978). Most of the information to support these interpretations has been obtained from sediments and fossils sampled on the floors of the ocean basins. On the continental areas each succeeding ice sheet has tended to destroy the evidence of the preceding interglacial and glacial environments. The cores from the ocean floor sediments (Sancetta, Imbrie & Kipp 1973; Cline & Hays 1976; Hays, Imbrie & Shackleton 1976; Kellogg 1976; Ruddiman & McIntyre 1976; Ruddiman, Sancetta & McIntyre 1977), reveal a sequence of glacials lasting about 100,000 years with interglacials persisting only for 10,000 years. Environments similar to those of the present, for so long thought to be typical of most of the Quaternary period appear to be representative of only 10 per cent of the time, the remainder of which, in contrast, was predominantly cold and possibly semi-arid in mid and high latitudes.

Little can be written with confidence about Quaternary environments in Scotland before 130,000 bp. It is highly likely that several ice sheets developed and wasted away before that date and that there were long periods of intense periglacial activity associated with tundra environments. It is, therefore, not at all surprising that all the evidence of Tertiary and Early and Middle Quaternary environments has been removed except for the much modified (by glacial and periglacial processes) landform assemblage which was largely developed during the Tertiary period.

There are two important sources of evidence outwith Scotland which provide some indication of the environmental conditions which prevailed within Scotland during the last 130,000 years. Firstly, the analysis of deep oceanic sediments in the north-east Atlantic Ocean (Sancetta *et al* 1973) provide data on sea surface temperatures (Fig. 1.9) which can be used to determine probable temperature trends in Scotland. Secondly, Coope (1977) has constructed a temperature curve for southern and central England during the last 120,000 years which can also be used in an analysis of the Scottish environment over the same period.

TABLE 2.1

Estimated sea surface temperatures at core site V23-82 (Sancetta et al 1973) *and estimated air temperatures at sea level and at 600 m in Scotland during the last 127,000 years.*

Thousand years bp	Sea surface temps V23–82 °C		Sea surface temps W. coast Scotland		Sea level air temps W. coast Scotland		Air temp at 600 m	
	JAN	JULY	JAN	JULY	JAN	JULY	JAN	JULY
0	10	15	8	12	4	14	0	10
10	5	13	3	10	−1	12	−5	8
11	7	13	5	10	1	12	−3	8
11·5–21·8	0	6	−2	3	−6	5	−10	1
31	4	11	2	8	−2	10	−6	6
36	0	7	−2	4	−6	6	−10	2
48	7	12	5	9	1	11	−3	7
51	0	8	−2	5	−6	7	−10	3
59	4	10	2	7	−2	9	−6	5
73	0	6	−2	3	−6	5	−10	1
81	9	15	7	12	3	14	−1	10
109	2	9	0	6	−4	8	−8	4
124	9	16	8	13	4	15	0	11
127	1	8	−1	5	−5	7	−9	3

The present day sea surface temperatures at core site V23–82 (Sancetta *et al* 1973) some 1000 km south-west of Scotland (Fig. 1.8) are within two or three degrees centigrade of the sea surface temperatures which occur off the west coast of Scotland today (Fig. 1.1). It would not seem unreasonable to use the temperature curves (Fig. 1.9) derived from that core as analogues for sea level temperature conditions in Scotland during the same period with one proviso. During the cold

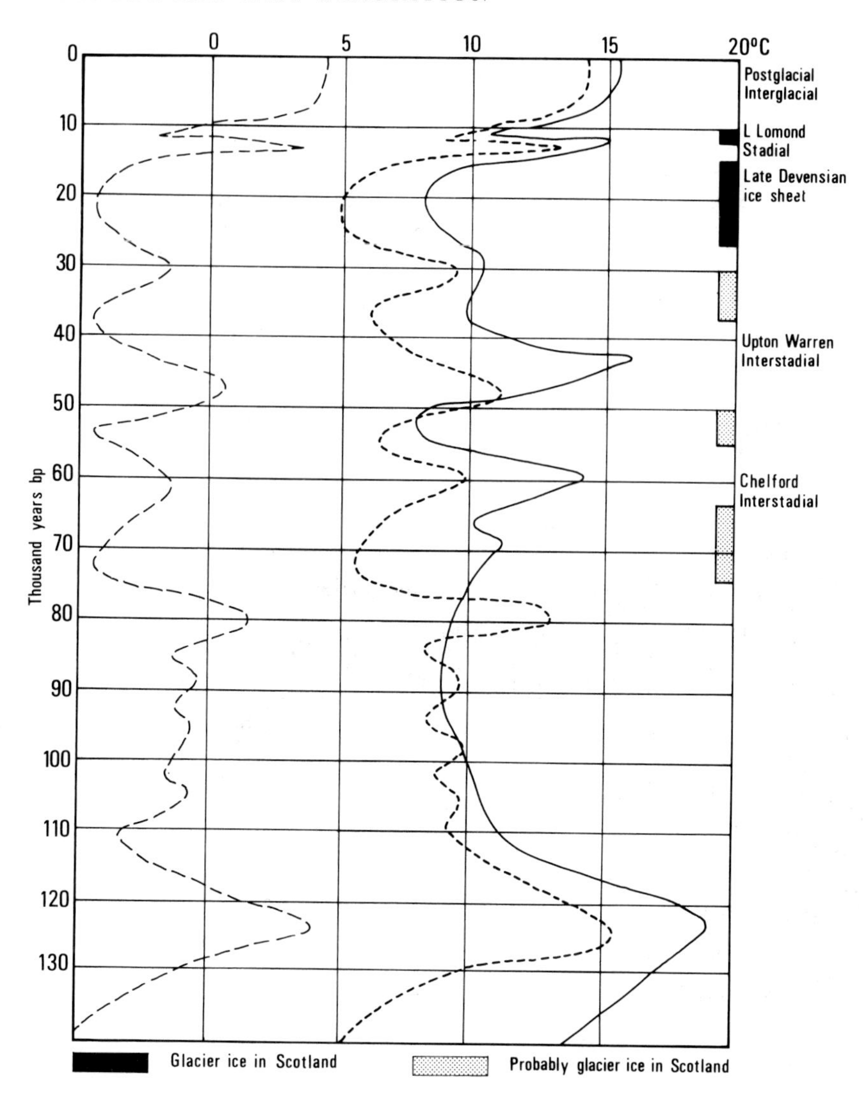

Fig. 2.1. Estimated sea-level air temperatures—January, July (dashed lines)—for west central Scotland and Coope's (1977) estimated summer temperatures in lowland central Britain based on the evidence of fossil *Coleoptera*.

episodes at the core site it is likely that certainly winter temperatures and probably summer temperatures in Scotland would have been several degrees lower because of the increased continentality of the Scottish climate at those times, and as a result of the presence of sea-ice around the Scottish coasts in winter. Work by Kellogg (1978; 1980) on the palaeoclimatology of the Norwegian Sea (the area between Iceland, Scotland and Norway) has indicated that there was a permanent sea-ice cover off the north-west coast of Scotland at 18,000 bp and 82,000 bp and that sea surface temperatures at 120,000 bp were similar to those of today.

It is possible to construct a table (Table 2.1) using the estimated sea surface temperature at core site V23–82 and subtracting 2°C and 3°C respectively for January and July to obtain estimated sea surface temperatures on the west coast of Scotland. By subtracting 4°C from the January figures and adding 2°C to the July figures it is possible to produce January and July mean air temperatures at sea level for western Scotland. By using a lapse rate of 0·65°C per 100 m it is also possible to estimate mean air temperatures at 600 m in January and July.

It would indeed be unwise to place too much faith in the absolute temperature values estimated for sea level and at 300 m above sea level in western Scotland. Specifically, the presence of glaciers and/or an ice sheet in Scotland at any particular time would depress the estimated values considerably. It is probably better to use the estimated air temperatures as indicative of temperature trends and with this in mind they have been plotted on a graph (Fig. 2.1). To this graph has been added a temperature curve produced by Coope (1977) for central England.

Coope (1977) has shown that abundant fossil Coleoptera to be found in the deposits of Devensian age in England reveal (p. 313), "marked changes in the geographical distribution of Coleoptera during the last glacial–interglacial cycle [which] conform to an orderly pattern of climatic fluctuations." He argues that, since Coleoptera display a remarkable degree of evolutionary stability, and that communities of species are drawn together by common ecological preferences which are well-established by studies of their modern distribution patterns, they can be used with confidence to reconstruct the past environments of the deposits in which the fossils occur. In particular Coleoptera are especially valuable as indicators of temperature, and on that basis Coope has produced a mean July temperature curve for central Britain during the last 120,000 years (Fig. 2.1).

The close similarity in trend between Coope's curve for mean July temperatures and the July curve for western Scotland based on the data from core V23–82 (Fig. 2.1) is quite remarkable. At present, mean July sea level temperatures in west central Scotland are 1°C cooler than those of central England, and the differences between the estimated temperatures over the last 130,000 years are mainly less than 3°C. However, at altitudes above 300 m the Scottish mean January and July temperatures are 2 to 3°C cooler than those of Birmingham. At even higher altitudes (over 1000 m) the differences are much more significant (4–6°C).

Bearing the above comments in mind and stressing that comparisons with the English Midlands should only be made in terms of those areas in Scotland lower than 300 m above sea level, the data now available for the English Midlands may be used to give some indication of environmental conditions in Scotland during the Ipswichian Interglacial and the Early and Middle Devensian intervals.

During the Ipswichian Interglacial mean July temperature was higher than at present in lowland central Britain. About 120,000 bp the climate had begun to cool and between 105,000 bp and 75,000 bp mean July temperature was 9°C. During the Chelford Interstadial, about 60,000 bp, the climate was rather cooler than now

(July mean temperature: 14°C) with a moderate degree of continentality. The fossil insect assemblage at Chelford indicates a climatic regime similar to that in southern Finland today between 60° and 65°N latitude. The botanical evidence from Chelford (Simpson & West 1958; Coope, Shotton & Strachan 1961) indicates the establishment of forest, consisting of pine, birch and spruce, during this interstadial. Although the summer temperatures may have been high, it is likely that winter temperatures remained low. In regions where pine, birch and spruce are typical to-day mean January temperatures in the range −10°C to −15°C are not uncommon.

About 48,000 bp there was another very sharp rise and fall in July mean temperature—the Upton Warren Interstadial. Coope (1977, p. 322) suggests that "Before the thermal maximum of the interstadial . . . the climate was of arctic severity and that within one thousand years mean July temperatures rose by 8°C." The fall in July temperatures after the thermal maximum may not have been as rapid but the period 40,000 to 30,000 bp was characterised by July temperatures at or just below 10°C with exceedingly cold winters (mean January temperatures below −20°C). The Upton Warren thermal maximum was so short that there was not sufficient time for trees to re-colonise central England and the vegetation of this period is characterised by an absence of trees and a dominance of herbs.

It would seem likely therefore, that for much of the period 130,000–30,000 bp Scotland would have experienced a cold continental climate with a treeless, tundra vegetation. Only for two short periods (62,000–55,000 bp and 46,000–40,000 bp) did climatic amelioration take place. Only in the first of these interstadials is it probable that a forest cover was established.

One of the major characteristics of all three temperature curves shown on Figure 2.1 is the rapidity of temperature changes at various times during the last 130,000 years. The explanation of these remarkable temperature fluctuations appears to lie in the temperature of the oceanic waters lying to the west of the British Isles. The study of the North Atlantic oceanic sediments and their included foraminifera (Ruddiman & McIntyre 1973, 1976; Sancetta *et al* 1973; Ruddiman *et al* 1977) reveal that changes in sea surface temperatures of 7–11°C have occurred in a few thousand years. The major question then arises as to when did temperatures first descend low enough, after the Ipswichian Interglacial, to allow a major ice sheet to develop in Scotland. It is generally accepted that the last major ice sheet (the Late Devensian) to exist in Britain began to develop after 30,000 bp and had wasted away by 13,000 bp. Estimated temperatures similar to those of the Late Devensian glacial period on Figure 2.1 also occurred in the periods 35,000–40,000 bp, 52,000–55,000 bp, and 65,000–72,000 bp. If it is assumed that it takes about 8,000 years for a major ice sheet to develop in Scotland and to extend south to the English Midlands, it seems unlikely that an ice sheet of the dimensions of the Late Devensian ice sheet was established in any of these periods. However, since it is known that major glaciers over 50 km long developed in the Lateglacial (Loch Lomond) Stadial in Scotland in less than 1,000 years (11,000–10,000 bp), it is likely that at least valley glaciers existed in Scotland at 35,000 bp, 55,000 bp and

70,000 bp. It should be noted that it was not until after 75,000 bp that January sea surface temperatures at the site of core V23–82 (Table 2.1) began to fall below 1°C and Ruddiman *et al* (1978) have suggested that although at least 75 per cent of the total ice expansion of the last glacial episode had been achieved by 75,000 bp there was, at that time, marked oceanic warmth in the North Atlantic and sea surface temperatures off the west coast of Scotland were similar to those of to-day.

On the basis of Figure 2.1 a tentative summary of environmental conditions in Scotland between 130,000 and 30,000 bp can be made:

125,000–115,000 bp: Temperatures at least as warm as those of to-day. An Interglacial environment (Ipswichian) with the establishment of an oak/pine forest with relative sea level as high and possibly higher than that of the present.

115,000–82,000 bp: Temperatures generally 5°C lower than at present with short periods (circa 1,000 years) with truly periglacial conditions. Relative sea level falling as the northern hemisphere ice sheets expanded. There is a possibility that small valley glaciers could have developed from time to time in the Scottish mountains.

82,000–78,000 bp: An interstadial during which temperatures were similar to those of to-day.

78,000–65,000 bp: A glacial/periglacial environment probably with at least a major glacier complex if not an ice sheet in Scotland. A probable rise in relative sea level due to isostatic depression due to glacier loading.

65,000–55,000 bp: A major interstadial (Chelford) with temperatures about 5°C lower than to-day, but with the possibility of the establishment of a forest cover even though mean January temperatures were below 0°C. Relative sea level probably falling.

55,000–50,000 bp: A return to a glacial/periglacial environment with valley glaciers in the Scottish mountains.

50,000–45,000 bp: An interstadial (Upton Warren) with temperatures some 2–4°C lower than at present. Probably a falling relative sea level.

45,000–32,000 bp: A return to a glacial/periglacial environment with the build up of a major glacier complex and even possibly an ice sheet in the Scottish mountains.

32,000–28,000 bp: An interstadial with temperatures about 5°C lower than those of to-day.

It must be stressed that the chronology of the environmental changes described above is highly suspect. However, it can be stated with some confidence that both the magnitude and frequency of environmental changes between 130,000 and 280,000 bp in Scotland were considerable. It is now necessary to examine the available evidence obtained from the sediments and their included fossils which were associated with these various environments.

Sedimentary sequences incorporating organic remains suitable for radiocarbon dating and/or pollen analysis which are older than the deposits laid down by the last Scottish ice sheet (Late Devensian) are rare. Only recently has the first site containing a sequence of glacial tills separated by palaeosols been described at

Kirkhill in Buchan, north-east Scotland (Caseldine & Edwards 1982, and Connell *et al* 1982). This site demonstrates that the Buchan plateau has been glaciated on at least three separate occasions. The stratigraphy at the site (Fig. 2.2) involves some 2·3 m of sands and gravel on top of which there is a podzol-like horizon from which material was obtained for radiocarbon dating. The dates obtained were: 45,630+1740–1430 bp; 44,900+1580–1320 bp; and 33,810 + 630–590 bp. Above this podzol there were 1–2 m of solifluction deposit overlain by 1–2 m of weathered till which in turn were overlain by 0·5 m of sandy solifluction deposit on top of which there were 1·5 m of unweathered till. Connell *et al* (1982, p. 571) interpret the sequence of events represented by the stratigraphy at Kirkhill as follows:

"1. Erosion of channels/basins and deposition of fluvial or glacifluvial sands and gravels.
2. Weathering and pedogenesis under humid temperate interglacial conditions to produce a podzol-like soil profile in the sands and gravels.
3. Disturbance and burial of the podzolic profile by periglacial sediments followed by deposition of till.
4. Weathering and soil development in the till under a second period of humid temperate climate.

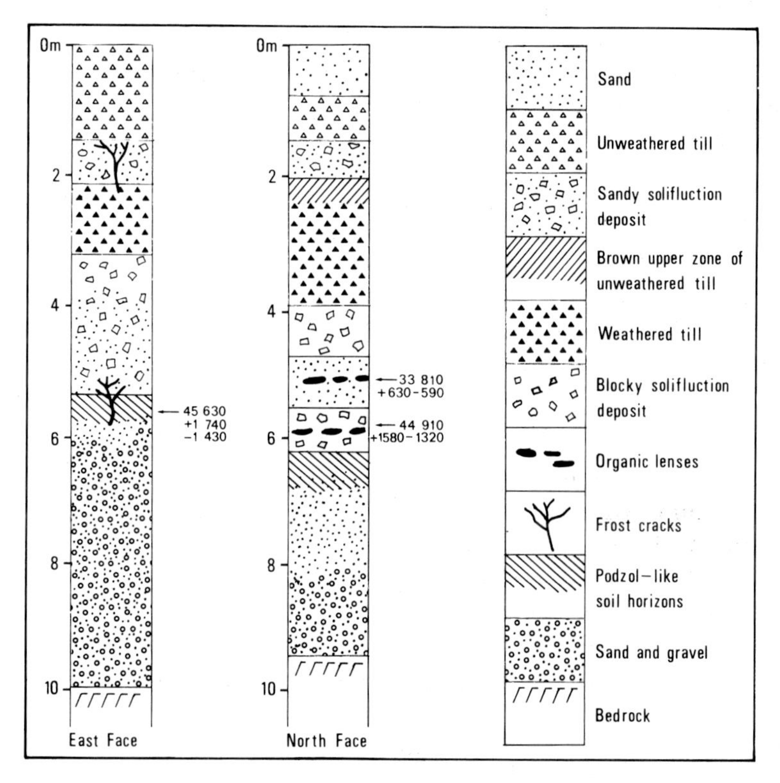

Fig. 2.2. The stratigraphy at Kirkhill Quarry, Buchan (Connell *et al.* 1982).

5. Periglacial disturbance of the soil profile and final glaciation of the site resulting in the deposition of the fresh, unweathered, upper tills.
6. Erosion of channel and deposition of glacifluvial sands (north face).
7. Flandrian pedogenesis in tills and sands."

The authors reject the radiocarbon dates of Middle Devensian age since they are associated with pollen of a relatively thermophilous nature and the dates are regarded as too young as a result of contamination. However, in the context of the estimated temperature on Figure 2.1, perhaps these dates should not be rejected out of hand in that they could possibly represent an interstadial (cf. Upton Warren) between 40,000 and 50,000 bp.

On the basis of counting cold/warm strata from the top of the sequence downwards, Connell *et al* (1982) prefer to equate the lower weathered till which contains pollen of Alnus (73·9 per cent of the total land pollen sum) and Betula (4·3 per cent) and coryloid grains (8·7 per cent) with the Ipswichian Interglacial and the Gramineae-Alnus-Solidago assemblage of the lower palaeosol with the Hoxnian. However, they go so far as to suggest that the upper palaeosol could be Hoxnian and the lower one Cromerian in age.

The problems of interpretation of the Kirkhill sequence are typical of those associated with all the pre Late Devensian (i.e. earlier than 30,000 bp) sites in Scotland. Radiocarbon dates are rejected on the assumption that they are too young because all organic material suggestive of relatively warm conditions must be at least Ipswichian in age. Perhaps the estimated temperature curves shown on Figure 2.1 may lead to a review of the evidence at Kirkhill and some of the other sites discussed below.

At the time of writing there are 17 radiocarbon age determinations (Fig. 2.3) on materials from Scotland (including the continental shelf) which are older than 30,000 bp, of which 11 are regarded as infinite. None of these infinite dates adds anything to an interpretation of the chronology of environmental change in Scotland. However, the peats sampled at the Burn of Benholm and at Fugla Ness do give an indication, as a result of pollen analysis, of the vegetation which existed in those areas at the time of peat formation. At Fugla Ness (Birks & Ransom 1969) in Shetland over 1 m of peat is covered by and rests upon glacial till. The contents of the peat indicate the former existence of open coniferous woodland, alternating with heath and grassland communities. The authors suggest a climate warmer than that of to-day with mild frost-free winters and the organic deposits are tentatively correlated with the Hoxnian Interglacial. However, a series of radiocarbon dates was obtained from the Fugla Ness site and a similar site at Walls by Page (1972). The dates range between 36,000 and 40,000 bp. Along with other anomalous dates obtained by Page for some English interglacial deposits (Shotton 1972), the dates from Shetland are generally considered to reflect contamination and are therefore rejected.

Pollen analysis of the peat below a glacial till at the Burn of Benholm (Donner 1979) again reveals a tundra vegetation dominated by grasses and herbs. However, with an infinite radiocarbon date little more can be said than that these tundra conditions existed prior to 42,000 bp.

The site at Kilmaurs in Ayrshire consisted (it is no longer visible) of 10 m of stratified deposits, which included organic remains, overlain by more than 20 m of glacial till. Fossils obtained from these deposits include: mammoth tusks and teeth, reindeer antlers, beetles, plant remains and molluscs indicative of arctic and subarctic conditions. Again the infinite date on the reindeer antler (Shotton *et al* 1970) is of little help in indicating the chronology of the environments indicated by these deposits and their included fossils.

The three sites at Teindland (Fitzpatrick 1965; Edwards *et al* 1976), Bishopbriggs (Rolfe 1966) and Tolsta Head (Von Weyman & Edwards 1973) appear to be

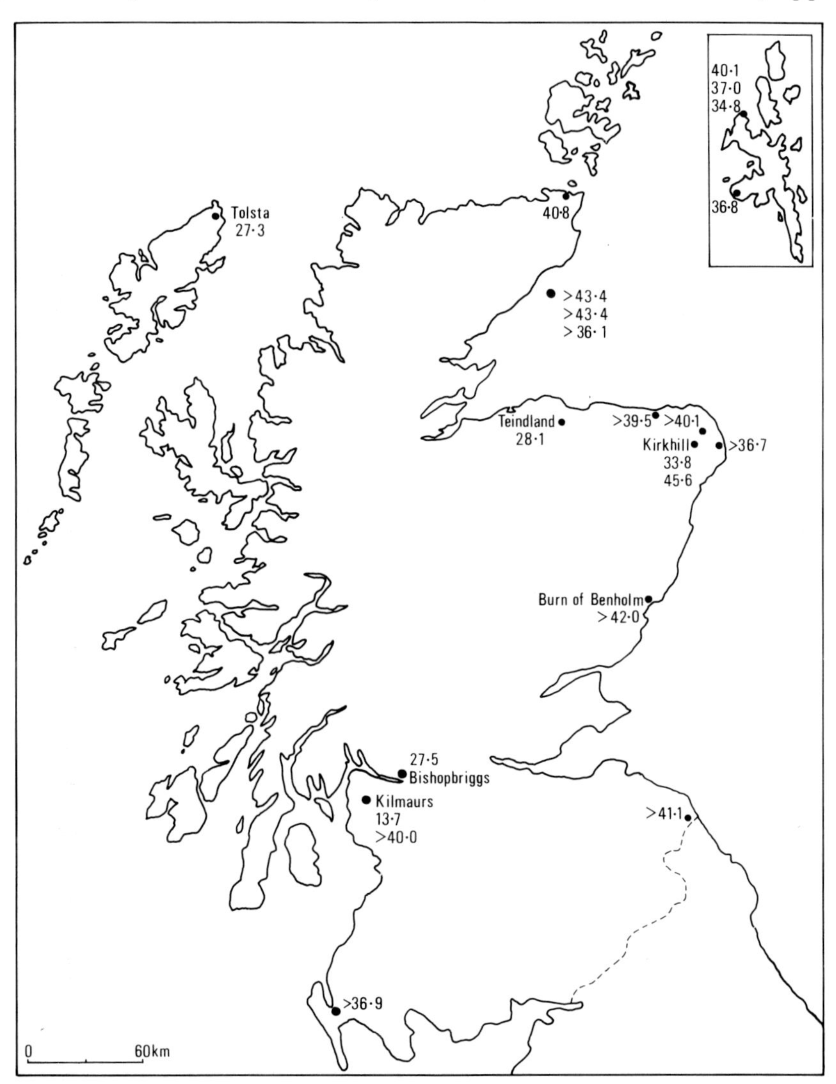

Fig. 2.3. Sites in Scotland from which radiocarbon dates older than 27,000 bp have been obtained.

much more reliable in the information they provide about the environment in Scotland immediately prior to the onset of the last glaciation.

The Teindland site provided a radiocarbon date of 28,100 + 480, −450 bp for a fossil soil covered by 2 m of material interpreted as till and outwash gravels by Fitzpatrick (1965). Romans *et al* (1966) suggested that the material overlying the dated soil is a solifluction deposit and implied that this locality, along with a large area in north-east Scotland, was not covered by the last ice sheet. Edwards *et al* (1976) have re-examined the Teindland site and conclude that (p. 742) ". . .the fossil microflora contained within the soil can be assigned to two distinct phases of geological time, the Middle Devensian Interstadial and the Ipswichian stage." It is the so-called Middle Devensian material which is of interest here. The radiocarbon date of 28,140 + 480, −450 bp is identical with that obtained by Fitzpatrick in 1965. Edwards *et al* (1976) point out that the vegetation associated with this buried podzol (*Pinus sylvestris, Calluna vulgaris* and *Gramineae* spp) may indicate a retrogressive vegetation succession. Immediately beneath the till an organic horizon provided 97·6 per cent herbaceous pollen (mainly *Gramineae* and *Rubiaceae* spp) which they believe may indicate the cold climatic conditions similar to the more oceanic conditions deduced at Tolsta Head (Von Weyman & Edwards 1973), for the same period.

Throughout the nineteenth century numerous specimens of mammalian bones, teeth, tusks and antlers (Pl. 2.1A, B) were recovered from the gravel pits in central Scotland. With the advent of mechanised extraction of gravel very few such specimens have been recovered in recent years. However, in 1963 a left humerus of a woolly rhinoceros (*Coelondonta antiquitatis*) was picked off a conveyor belt at the Wilderness pit, Bishopbriggs (Rolfe 1966). Since the bone was not found *in situ* but was dug from the working face by a mechanical excavator the exact horizon is unknown. However, after consultation with the pit operators Rolfe was confident that the sand horizon in which the bone occurred was overlain by a reddish-brown, sandy clay with pebbles, believed to be the till deposited by the last ice sheet to cover the area. The top of the ossiferous sand was penetrated by a clay-filled fissure which Rolfe (1966, p. 255) interpreted as ". . .one of the numerous ice-wedge casts recently recognised in the pits of the area."

About 150 g of the bone from the Wilderness pit were used for radiocarbon dating and yielded abundant and well-preserved collagen. The collagen was dated at 27,550 (+1370, −1680) bp. It therefore appears that the Kelvin Valley, north-east of Glasgow, had a periglacial environment about 27,000 bp in which woolly rhinoceros grazed a tundra type vegetation. Some time after that date the arrival of the glaciers, which were responsible for the fluvioglacial deposits in which the remains of the woolly rhinoceros were found, presumably led to the extinction of the species in this part of the world.

At Tolsta Head on the Island of Lewis, Von Weyman & Edwards (1973) obtained a date of 27,333, ±240 bp from organic detritus in laminated lake-silts beneath a glacial till. They state (p. 473), "The high frequencies of grass pollen and light demanding herbaceous species such as *Calluna vulgaris, Compositae* and Ranunculaceae indicate an open vegetated landscape. . . The taxa present . . . do not necessarily indicate a severe stadial climate. . . The pollen spectra are not inconsistent with the cool maritime climate of Lewis. The increased representation of the warmth loving juniper in the upper section of the diagram might be considered

Plate 2.1A. Woolly Rhinoceros (*Coelondonta antiquitatis*) (Drawing based on a stuffed specimen in the Zoological Museum in Krakow).

Plate 2.1B. Woolly mammoth (*Mammathus primigenius*) (Kurten 1972).

anomalous before the impending glacial advance responsible for the overlying till. Although the species possess a wide ecological amplitude a possible interpretation is that the deposit has been truncated by glacial erosion. The profile therefore may be incomplete." The site at Tolsta Head, therefore, is a useful additional piece of evidence relating to the date of the last glaciation in Scotland but the palaeobotanical evidence seems to contain some contradictory evidence concerning climatic conditions prior to the onset of glaciation.

From the above discussion of the sites in Scotland from which radiocarbon dates older than 27,000 bp have been obtained it can be seen that none of them is free of problems of interpretation. Assuming that the dates from Teindland, Bishopbriggs and Tolsta Head are reliable, then it can be stated that much of Scotland was not covered by glacier ice 27,000 years ago. However, there could have been valley glaciers in the higher ground at that time.

Combining Coope's (1977) evidence for central England, the palaeobotanical evidence from Teindland and Tolsta Head, the finds of woolly rhinoceros at Bishopbriggs, and mammoth tusks, teeth and reindeer antlers at Kilmaurs, it would appear that lowland Scotland had a severe periglacial environment 30,000–27,000 years ago. Mean January temperatures probably were of the order of −5°C and mean July temperatures of the order of 5°C. The vegetation was dominated by grasses, shrubs and herbs with some open woodland. In this tundra-like environment the woolly mammoth (*Mammathus primigenius*), the woolly rhinoceros (*Coelodonta antiquitatis*) and the reindeer (*Rangifer tarandus*) were important members of a fauna which was to be wiped out by the ensuing glaciation. Only the reindeer would return to Scotland after deglaciation.

The woolly mammoth is almost a symbol of the Ice Age. We know a great deal about it because of the discovery of frozen specimens in the permafrost of Siberia and Alaska. Kurten (1972) states that, "The remarkable towering shape (shoulder height of 4 m) of the mammoth with its peaked head and humped, sloping back, its long hanging hair and the great extravagantly curved tusks, add up to an unforgettable apparition that was recorded time and time again by man in their paintings and engravings." These great animals grazed the steppes and tundras of Europe during the waxing and waning of the Pleistocene ice sheets.

A specimen of the woolly rhinoceros (*Coelodonta antiquitatis*) has also been found at Starunia, at the foot of the Carpathian mountains, in a 13 m thick layer of silt imbued with salt and petroleum vapour. There is a complete stuffed specimen in the Museum at Krakow. This animal also appears in Stone Age paintings and engravings. According to Kurten (1968, p. 143), "The woolly rhinoceros is perhaps mostly thought of as an extreme tundra form, but unmistakeable specimens of this species have also been discovered in completely different surroundings . . . for example . . . near Barcelona in deposits with pollen indicating a dry temperate climate with extensive grassland and a few broad leaved trees." However, the woolly rhinoceros occupying Scotland 27,000 years bp almost certainly occupied a periglacial environment with permafrost in the ground. The expansion of glacier ice after 27,000 bp presumably drove these animals and the mammoth and reindeer southwards.

The position of the Scottish coastline during the Quaternary period varied considerably partly as a result of eustatic fluctuations of sea level reflecting the expansion and retreat of the great ice sheets and partly as a result of local isostatic adjustments. Evidence relating to the exact position of the coastline and/or the relative altitude of shorelines at specific dates prior to 14,000 bp is not available. There is evidence of a high-level stand of the sea (up to 43 m above present sea level) in western Scotland prior to the onset of the last glaciation in the form of a high-level rock platform and cliffline with associated sea stacks which bears evidence of being modified by glacier ice. Wright (1911) identified traces of this 'pre-glacial' shoreline in the Inner Hebrides at altitudes of between 27 and 41 m above present sea level. It is now widely accepted that these features are more likely to be early Devensian or pre-Devensian rather than pre-Quaternary in age (McCann 1964; Sissons 1974d) and that they may have been produced during one or more interglacial intervals.

This high-level rock platform has been identified from Islay in the south to the Applecross peninsula in the north (Cunningham-Craig *et al* 1911; Bailey *et al* 1924; McCann 1964; 168; McCann & Richards 1969; Richards 1969; Jardine 1977). The platform is believed to be warped (McCann 1968, p. 26), declining in height both to the north and to the south away from a centre in Ardnamurchan. It may also be a composite feature consisting of at least two features as identified on Oronsay by Jardine (1977, p. 103). The date of formation of these high-level platforms is unknown and Jardine points out (1977, pp. 103–104), ". . .the evidence from Oronsay suggests a much more complex history of shoreline development associated with the 'high-level platform' of the Inner Hebrides than hitherto has been recognised."

High level deposits (between 34 and 145 m above present sea level) containing marine shells have been known for over a century in Scotland (Sutherland 1981). There has been considerable debate as to whether these deposits are *in situ* or have been emplaced by glacial rafting. Sutherland (1981) argues that they are *in situ* and that they represent a marine invasion of the coastal areas due to isostatic down-warping during the advance of the last (Late Devensian) ice sheet. All the dates on shells from these deposits are infinite and greater than 39,500 bp. As Sutherland (1981, p. 251) states, "These dates apparently present a problem for the interpretation of the shell beds offered in this [his] paper, for it is generally argued that the last Scottish ice sheet advanced after circa 27,000 years bp (Sissons 1976). The infinite radiocarbon dates suggest that the shell beds are considerably older than the most recent advance of the Scottish ice sheet and hence could not be the result of crustal downwarping during this advance." Both Sutherland (1981) and Sissons (1981b) explain away this difficulty by suggesting that the last Scottish ice sheet developed during the Early Devensian (circa 75,000 bp). Until such time as finite radiocarbon dates are obtained on deposits between 75,000 and 30,000 years old, such a conclusion is unsupportable.

The stratigraphic, palaeontological, palynological, geomorphological and radiocarbon evidence relating to environmental changes in Scotland between 130,000 and 30,000 bp is inconclusive. It is suggested that the evidence obtained from oceanic cores in the north-east Atlantic Ocean and from the study of Coleoptera in the English Midland (Fig. 2.1) is more reliable in indicating the possible character of Scotland's environment during this period. Debates about the significance of 'infinite' radiocarbon dates (Sissons 1981a, 1981b; Sutherland 1981; Caseldine & Edwards 1982) and of undated palaeosols, marine deposits and pollen diagrams older than the last ice sheet clearly suggest that our knowledge of Scottish Ipswichian, Early and Middle Devensian environments is minimal. It is highly likely, on the basis of evidence from outwith Scotland that numerous and rapid environmental changes took place during these periods but until the deposits associated with those changes are accurately dated little more can be written. It seems likely that in Scotland the period 110,000–30,000 bp was dominated by periglacial and glacial environments with several short interstadials which may or may not have allowed the establishment of a forest cover. The presence of glacier ice at any particular time prior to 30,000 bp has yet to be determined.

The impact of the last ice sheet, which on the basis of the radiocarbon dates from Bishopbriggs, Tolsta Head, and Teindland did not develop until after circa 27,000 bp, was such that it either buried earlier deposits or largely removed them. For this reason the remainder of this book is concerned with Quaternary environmental changes associated with the build up, impact of and wastage of the Late Devensian ice sheet and the establishment of the Postglacial (Interglacial?) environment which ensued.

3

THE LAST ICE SHEET
circa 27,000–14,000 bp

THE NATURE OF THE EVIDENCE

By analogy with the deposits and landforms observed to be in the process of formation in association with existing ice sheets and glaciers it can be deduced that, at some time in the recent geological past, Scotland was covered by glacier ice. This deduction was first made one hundred and forty years ago (Agassiz 1840, A. Geikie 1863) and by 1865 Geikie was able to produce a map showing the principal directions of movement within the ice sheet (Fig. 3.1a). In 1901 A. Geikie published a map (Fig. 3.1b) on which the general lines of movement, the principal source

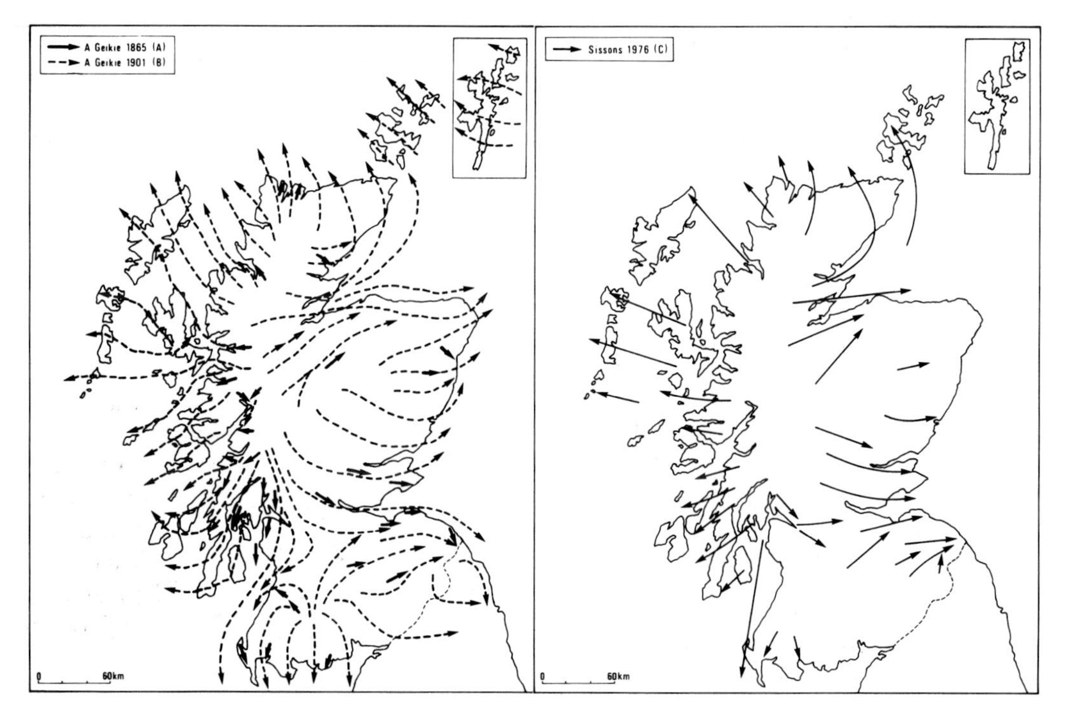

Fig. 3.1. Lines of movement in the Scottish ice-sheet as depicted on maps published by A. Geikie in 1865 (A) and 1901 (B) and J. B. Sissons 1976 (C).

areas and the main zones of conflict between ice originating in the Scottish Highlands and Southern Uplands were clearly identified. The map shows that the high ground in the western Highlands and western Southern Uplands was the principal source areas for the ice streams and that the entire area of the mainland was covered by the ice sheet. Ice originating on the mainland of Scotland flowed out beyond the Outer Hebrides and the Orkney Islands. Despite much detailed work being undertaken in the following seventy-five years little difference exists between Geikie's map and that produced by Sissons in 1976 (Fig. 3.1c). It will be shown below that although the former existence of an ice sheet covering Scotland is in little doubt, the extent of that ice sheet, the chronology of its build up and wastage and the environmental conditions associated with it are still poorly understood.

The evidence used to establish the former existence of an ice sheet in Scotland is primarily of two types. Firstly, the uppermost surface materials over much of Scotland consist of till or sand and gravel. These are deposits which, by analogy with areas of the world which presently contain glacier ice, can be demonstrated to have been laid down by glacier ice and meltwater. Secondly, Scotland contains many landforms which are known to be the product of erosional and depositional processes associated with glacier ice and meltwater. The depositional evidence is more reliable than the geomorphological evidence because, if sound stratigraphical principles are applied to the interpretation of the deposits, then at least the relative (if not the absolute) ages of the various sedimentary units can be determined. Over a very large part of Scotland the Quaternary stratigraphy is relatively simple and consists of one of four sequences: (a) glacial till resting on bedrock; (b) glacial till on bedrock and overlain by fluvioglacial sands and gravels; (c) bedrock overlain by fluvioglacial sand and gravels; (d) bedrock overlain by fluvioglacial sands and gravels overlain by till which in turn is overlain by fluvioglacial sands and gravels. All the sequences indicate only one period of glaciation. There are a few locations in which multiple till sequences are known to occur but they are very exceptional.

There can be little doubt, however, that Scotland has experienced more than one period of ice sheet glaciation during the Quaternary. On the basis of stratigraphic evidence from England it can be confidently stated that there have been at least three major ice sheets which overwhelmed Scotland during the last two million years and, on the basis of the recent climatic interpretations derived from the study of deep ocean sediments, it may be inferred that conditions conducive to ice sheet development in Scotland probably existed on many occasions during the Quaternary period. The existence of ice sheets on more than one occasion in Scotland is both the explanation of the relative simplicity of the glacial stratigraphy and the reason why the geomorphological evidence of the last glaciation is difficult to interpret. The last ice sheet was probably responsible for the removal of any glacial deposits laid down by its predecessors but at the same time it developed on a landscape which already had strong glacial characteristics created by earlier ice sheets. It is extremely difficult, for example, to establish the amount of erosion accomplished by the last Scottish ice sheet because the character of the land surface prior to the last glaciation is simply not known. If the concept of multiple glaciation is accepted, then the problem of the extent of inherited glacial characteristics becomes of considerable significance. Some of the large-scale features of glacial erosion—the glacial troughs and corries (cirques)—must have been modified by each successive ice sheet. Even relatively small scale features such as roches moutonnées or striae may not have been produced entirely by the last ice cover. Meltwater channels cut in bedrock may have been occupied during more than one period of ice cover. Such possibilities must inevitably make the interpretation of the geomorphological evidence rather difficult.

The evidence used to determine the lines of movement of the last ice sheet consists of glacial striae (scratches on rock surfaces), the alignment of streamlined erosional forms such as roches moutonnées (rock whale-backs), the alignment of glacial troughs and most significantly the occurrence of glacial erratics within and on the till deposits. The carry of both large and small rock fragments from a known outcrop to other localities can be clearly demonstrated in many areas of Scotland. The erratic train of the Lennoxtown essexite in the Central Valley (Peach 1909, Shakesby 1978) and the transportation of the distinctive Rannoch, Etive and Glen Fyne granites (Sissons 1967a, Fig. 30) are excellent examples of how ice flow lines can be established. It was evidence such as this, largely collected by the officers of the Scottish Geological Survey, which allowed James Geikie to construct his map of flowlines of the last ice sheet (Fig. 3.1b). Once these flowlines had been established then the main centres of ice accumulation could be deduced.

The question of the chronology of the build up of the last ice sheet in Scotland remains somewhat uncertain (see Chapter 2). All the evidence from southern Britain points to the total absence of glacier ice throughout the period 120,000 to 30,000 bp. However, this period was characterised by the build up of the Laurentide and Fennoscandian ice sheets and the study of deep-sea sediments indicates a world wide build up of glacier ice by 75,000 bp. Ruddiman *et al* (1980, p. 56) state that by 75,000 bp, "...the global ice volume reached 75–90 per cent of its maximum late Wisconsin value." They also state (p. 33), "During much of this rapid ice growth, the north Atlantic Ocean from at least 40°N to 60°N maintained warm sea-surface temperatures, within 1° to 2° of today's subpolar ocean." It was this relatively warm ocean that provided the moisture for the growth of the Laurentide and Fennoscandian ice sheets while at the same time providing climatic conditions which allowed the British Isles to remain ice free. If this assumption is correct then there are two important implications for the Scottish environment immediately prior to the onset of the last ice sheet. Firstly, sea-level would have been eustatically lowered by at least 120 m (Denton & Hughes 1981, p. 311) as the Laurentide and Fennoscandinavian ice sheets reached 75–90 per cent of their late Wisconsin maximum. Such a situation would have produced a remarkable change in the Scottish coastline (Fig. 3.2). Probably the most significant changes in the coastline were its migration westwards for 50–100 km and the large area of the North Sea basin which became dry land. Together these changes probably produced a much greater 'continental' element in the Scottish climate. Although the temperature of the North Atlantic Ocean to the west of Scotland during the summer was very little different from that

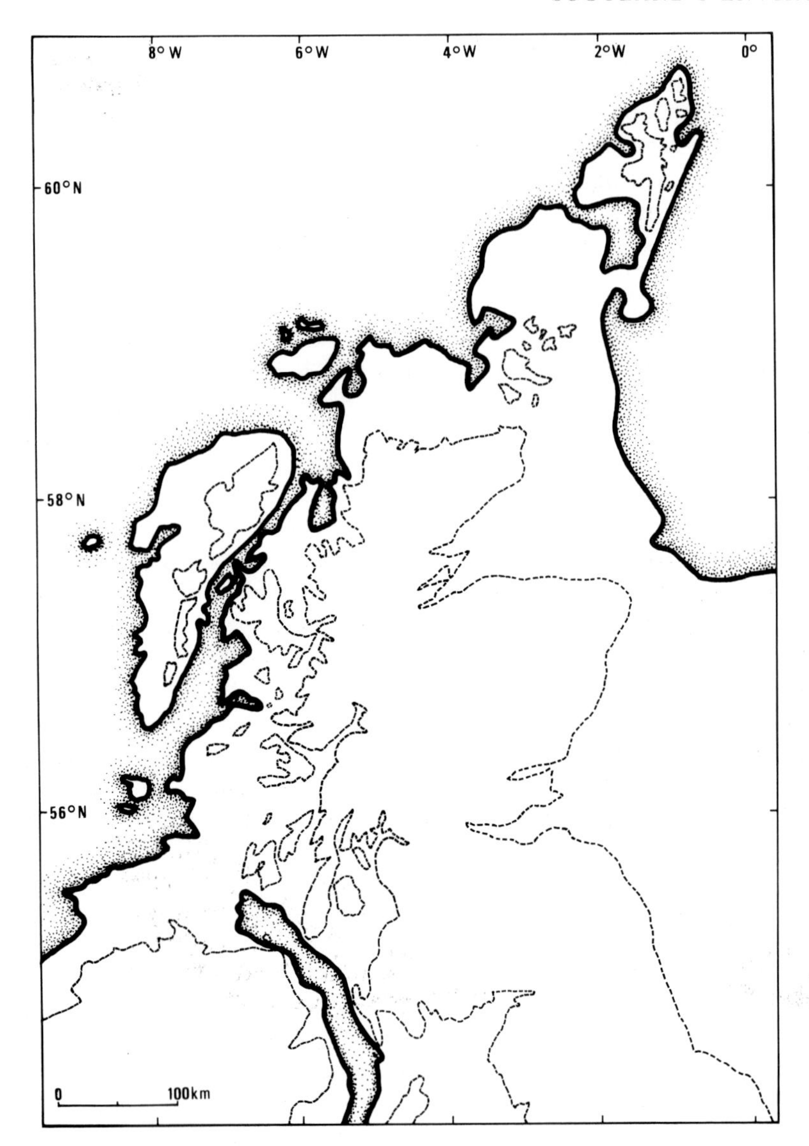

Fig. 3.2. The Scottish coastline when relative sea-level is reduced by 100 m.

of to-day, it is highly likely that with frontal activity being displaced northwards (Ruddiman *et al* 1980), the Scottish climate immediately prior to the onset of the last glaciation was considerably colder and drier than at present. The absence of any glacier ice in Scotland between 130,000 and 30,000 bp has often been questioned. It now seems likely that the lack of an ice sheet during this period is explained by the prevailing atmospheric conditions over the North Atlantic, but these do not preclude the possible development of valley glaciers (and glacier complexes) in Scotland at certain times (see Chapter 2) during the Middle Devensian (70,000–30,000 bp). When in Late Devensian times (i.e. after 30,000 bp) the polar front, separating polar and subpolar waters, began to move south, the storm systems bringing heavy snow falls migrated southwards and allowed snow accumulation and glacier growth in the western Highlands and western Southern Uplands.

MODELS OF THE LAST ICE SHEET

Three types of evidence have been used to create models of the last Scottish ice sheet. Geological and geomorphological evidence from the Scotttish mainland and islands allowed a model of the build up, lines of movement and extent of the ice sheet to be developed over one hundred years ago. This model established the main centres of accumulation in the western Highlands and western Southern Uplands and while establishing that all of the mainland plus the Outer Hebrides and Orkney Islands were overwhelmed by the ice sheet, it did not determine the extent of the invasion of Scottish ice into England, Wales and Ireland. Geological evidence, particularly the carry of Ailsa Craig microgranite into areas around the shores of the Irish Sea and of other Scottish erratics along the east coast of England extended the understanding of the Scottish ice sheet as a major contributor to the late Devensian British ice sheet. The third approach to modelling the last Scottish ice sheet is a very recent development. Using the geological evidence for flow lines and extent of the ice sheet along with formulae developed for theoretical ice sheet dynamics and making assumptions about accumulation and ablation rates, Boulton *et al* (1977) and Denton and Hughes (1981) have produced models which show the surface altitude and extent of the last Scottish ice sheet. Both models indicate that the main Scottish ice dome (Figs. 3.3, 3.4) reached an altitude of 1,780 m and that the ice was some 2,000 m thick over central Scotland. These models also indicate a secondary centre of ice dispersal in the central Southern Uplands. These reconstructions of the Scottish ice sheet suggest that it terminated to the west and north at the edge of the continental shelf and that the ice moved eastwards until it became confluent with Scandinavian ice. The value of these models is highly dependent upon the data fed into them and as Denton & Hughes (1981, p. 311) admit, "Our most important conclusion is that the distribution of late Wisconsin-Weichselian [Devensian] ice sheets is not well known." In the Scottish context Sissons (1981a, p. 16) echoes these sentiments when he states, "After 140 years of research it is surprising how little we really know about the extent of the last Scottish ice sheet and associated events." Bearing these comments in mind, the description of the last Scottish ice sheet which follows can only be regarded as tentative and will almost certainly have to be altered as new information about off-shore stratigraphy becomes available.

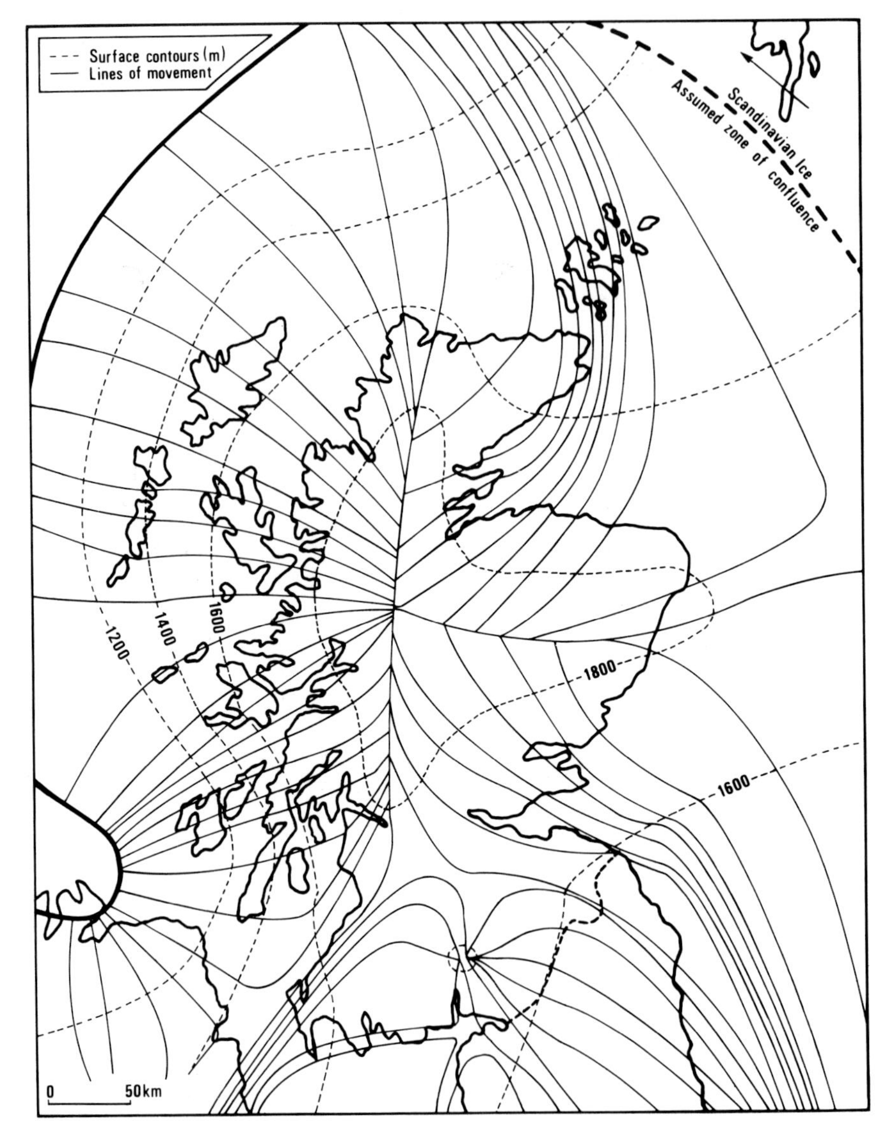

Fig. 3.3. The modelled surface topography and flow lines of the last Scottish (Devensian) ice sheet. (Boulton *et al.* 1977.)

THE BUILD UP, LINES OF MOVEMENT AND EXTENT OF THE ICE SHEET

The only detailed information about the extent of an ice mass which did not eventually completely overwhelm Scotland relates to the series of valley glaciers and relatively small ice caps which developed during the Loch Lomond Stadial (see

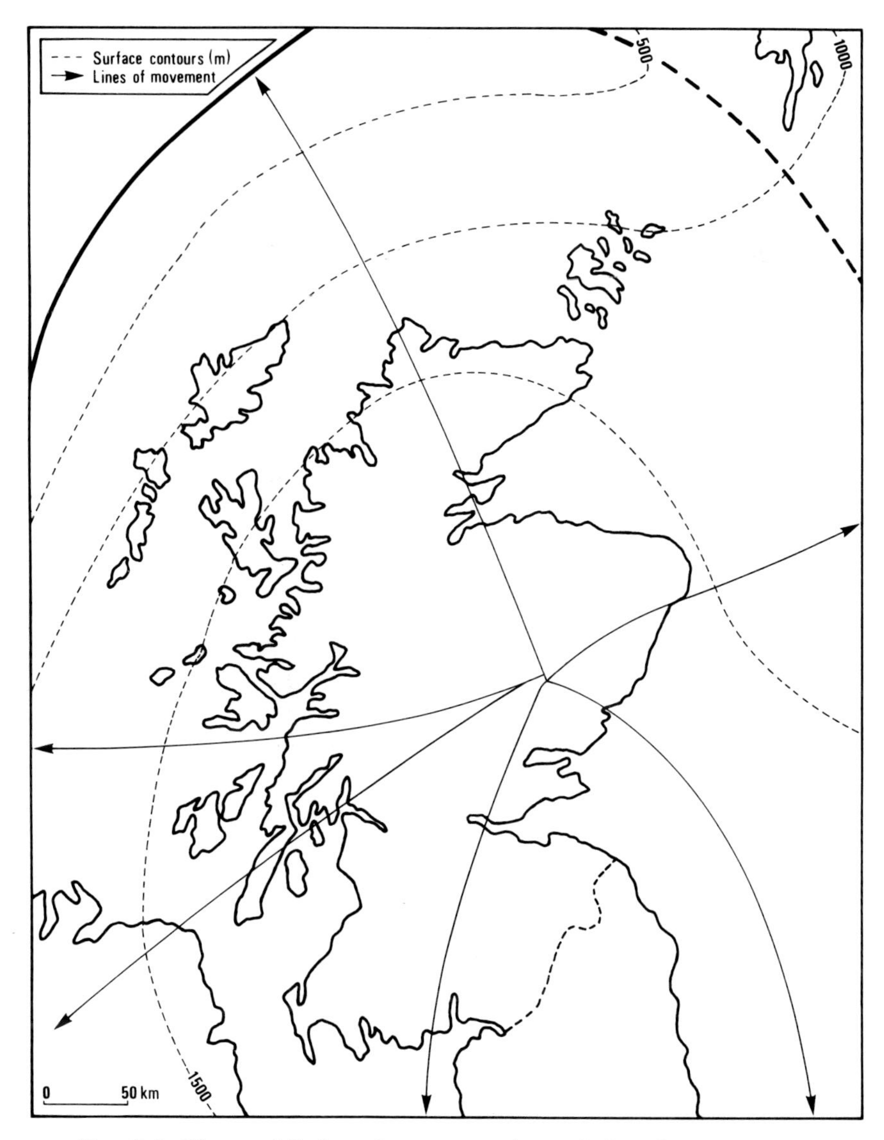

Fig. 3.4. The modelled surface topography and flow lines of the last Scottish (Devensian) ice sheet. (Denton & Hughes 1981.)

Chapter 4). Although the limits of the ice cover attained during that Stadial (Figs. 4.12 and 4.13) are now fairly well-known the time period over which the glaciers existed still remains uncertain but, on the basis of the radiocarbon dates currently available, it is likely that the period of glaciation lasted about 1,000 years. The origins and extent of the Loch Lomond glaciers may be regarded as analogous to the early phases of the establishment of the last ice sheet.

The deterioration of Scotland's climate resulting from the southward migration of the oceanic polar front in the North Atlantic was probably related to an increased frequency of atmospheric depressions passing across Scotland. The increased precipitation on the western mountains included larger winter snowfalls, and, with lower winter and summer temperatures, net accumulation of snow eventually exceeded ablation and glacier growth began. After about 1,000 years (Fig. 3.5A) individual glaciers may have attained lengths of up to 50 km and the glacier complex in the

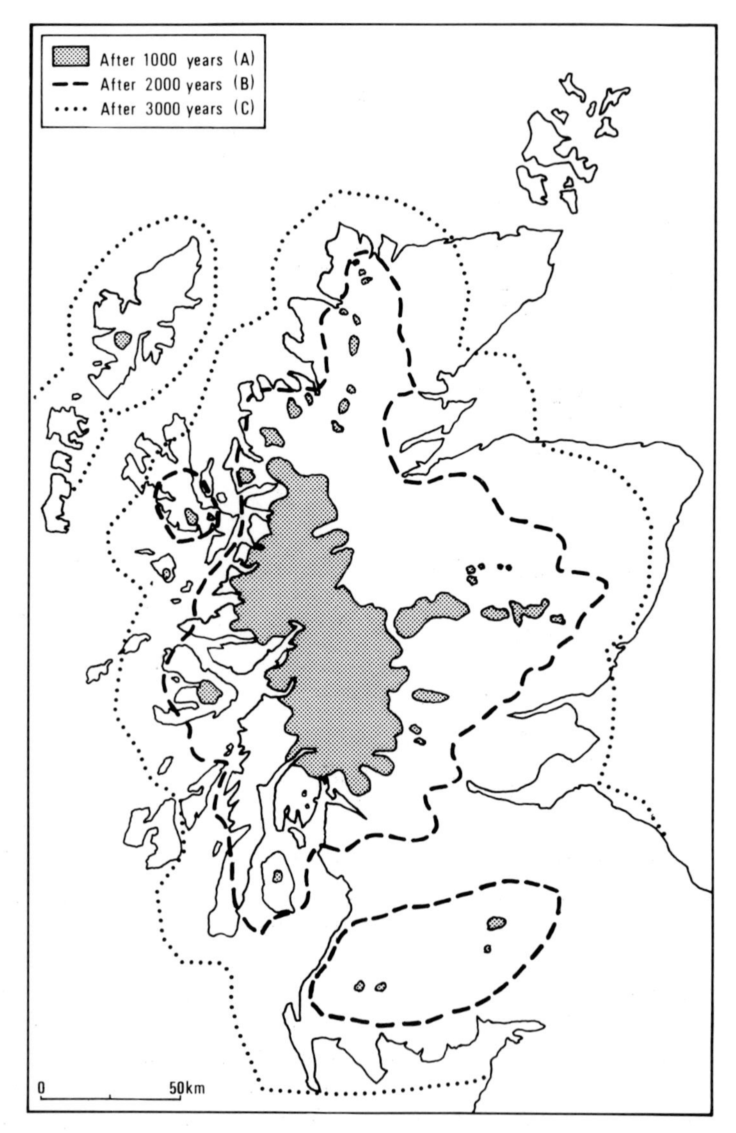

Fig. 3.5. Estimated ice margin positions during the build up of the last Scottish ice sheet.

western Highlands may have developed to form an ice cap. If it is assumed that climatic conditions during the initiation of the ice sheet were similar to those during the Loch Lomond Stadial, then only small glaciers (1–5 km long) existed at the end of the first thousand years of the glaciation in the Cairngorms and Southern Uplands. Small ice masses were also developing on Arran, Mull, Skye and Harris at this time. Beyond the margins of the glaciers a periglacial environment existed and

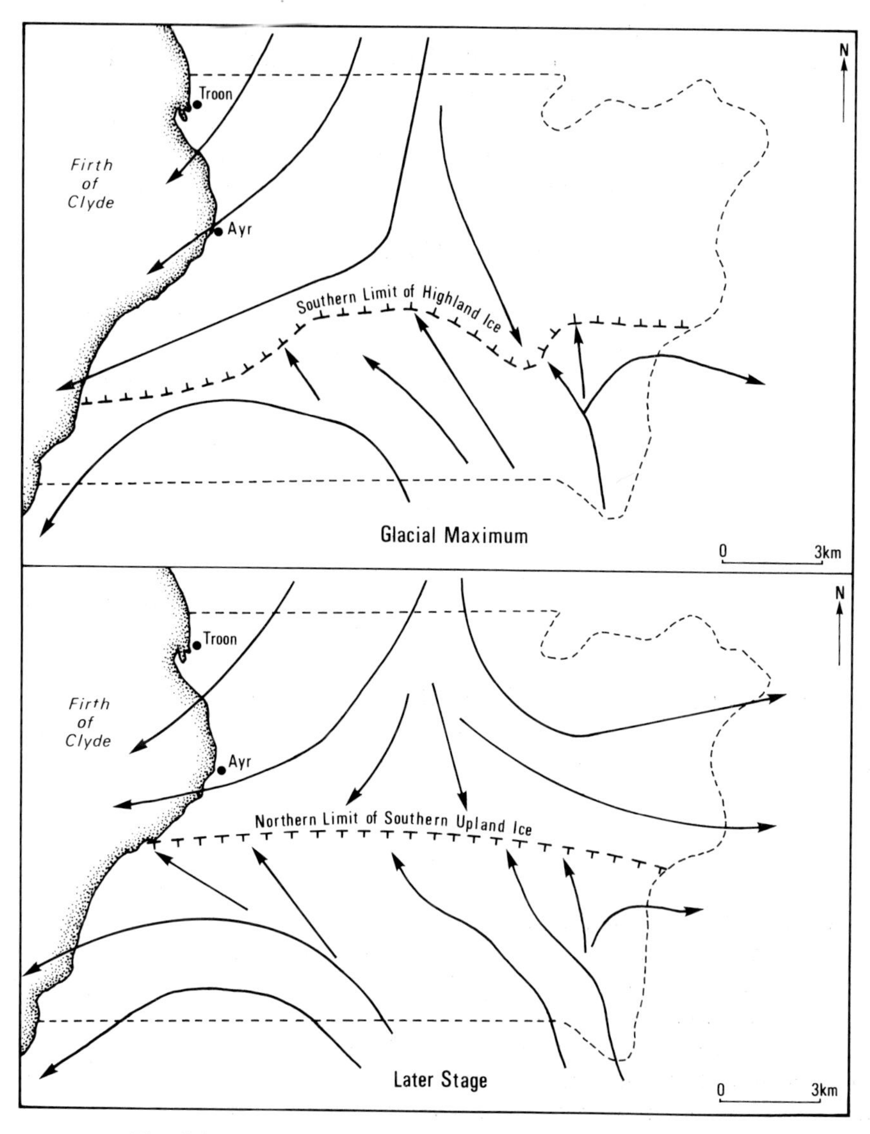

Fig. 3.6. Ice movements in Central Ayrshire (Holden 1977).

presumably with continued climatic deterioration, the plant and animal life of Scotland either migrated southwards or was progressively impoverished.

If it is assumed that continued climatic deterioration led to all the glaciers having positive mass balances then the frontal margins of each ice stream would advance. A probable rate of advance of an ice front in conditions of relatively high accumulation rates would be 30 m per year. After two thousand years of ice accumulation (Fig. 3.5B), all of the valleys in the western Highlands would contain major glaciers and ice streams would be moving down the Spey and Dee valleys, and being joined by tributary glaciers from the Cairngorms. The western Southern Uplands probably contained a major glacier complex but Southern Upland ice and Highland ice were not yet confluent in central Ayrshire and Lanarkshire. On the basis of this very simplistic model, Highland ice and Southern Upland ice would become confluent after about 2,500 years of glaciation. There is some stratigraphic evidence to suggest that Highland ice arrived in the so-called 'debatable ground' (A. Geikie 1887) in central Ayrshire before Southern Upland ice (Fig. 3.6). Holden (1977) has demonstrated that till with Highland erratics is overlain by till with Southern Upland erratics and he suggests that there was a relaxation of the pressure exerted by the Highland ice concurrent with an extension of Southern Upland ice. Sissons (1979e) has suggested that such an expansion of the Southern Upland ice mass and a halt or even contraction of the Highland ice mass could be explained by a migration of a zone of maximal snow fall associated with the southerly migration of the oceanic polar front, from the Highlands to the Southern Uplands, with relatively dry conditions occurring in the Highlands. There can be little doubt that the first two or three thousand years of the glaciation were dominated by the build up of ice in the western Highlands and ice streams radiated outwards in all directions. If a vertical thickening rate of 0·3 m per year is assumed for the ice mass in the western Highlands then, within 3,000 years (Fig. 3.5C) many of the drainage divides would be overtopped by glacier ice and the migration of the centre of ice dispersal to an area to the east of the principal north-south watershed would have taken place. Highland ice would have filled the Central Lowlands and become confluent with Southern Upland ice and although the Southern Uplands were an important ice accumulation area, it appears likely that there was a time lag which allowed Highland ice to occupy central Scotland before the nearer Southern Upland accumulation centre supplied large quantities of ice to that area.

It must be stressed that the sequence of stages described above is purely hypothetical as there is no direct evidence of the date and position of ice sheet margins during the growth phase. There is, however, some field evidence which suggests that the entire land surface of Scotland was eventually completely buried by the last ice sheet. Glacial striae have been observed at altitudes of up to 900 m in the northwest Highlands and glacial erratics occur at 1,100 m in Glen Coe, at 1,060 m on Schiehallion and on the tops of the Pentland, Ochil and Campsie Hills in Central Scotland. Such evidence indicates only minimum altitudes for the ice surface. It is also known that Scottish ice extended southwards into the Irish Sea basin and its known minimum thickness in North Wales requires a surface gradient which implies that the altitude of the source areas in the Scottish Highlands stood well above the hill tops.

Some authors have argued that the last Scottish ice sheet did not cover all of the Scottish mainland. Charlesworth (1955) suggested an ice-free area in NE Scotland and Fitzpatrick (1963, 1972) supported this contention. Detailed mapping of fluvioglacial landforms led Clapperton & Sugden (1977) to conclude that the whole of north-east Scotland was covered by the last ice sheet. Synge (1977) suggested that not only north-east Scotland but parts of Caithness, all of the Orkney Islands and northern Lewis were not covered by the last ice sheet. Sissons (1981a, p. 9) suggests that, ". . . none of Synge's supposed ice-free areas can be disproved at present." It is interesting that all these controversial areas are towards the periphery of the Scottish ice sheet. Until relatively recently it was generally accepted that ice originating on the Scottish mainland extended westwards beyond the Outer Hebrides and eastwards into the North Sea basin where it became confluent with Scandinavian ice (Fig. 3.1c) and was turned north-westwards over Caithness and Orkney, north-eastwards betwen Fife and Aberdeen and south-eastwards off Berwickshire.

Figure 3.7 shows the most recent interpretations of the lines of ice movement within the last Scottish ice sheet. The major changes in interpretation since 1976 (Fig. 3.1c) or for that matter since 1901 (Fig. 3.1a) are some minor alterations in the flow lines in Ayrshire (Holden 1977) and Dumfriesshire (May 1981), the reinstatement of an ice cover in north-east Scotland (Clapperton & Sugden 1977) and most significantly the establishment of independent centres of ice accumulation and dispersal in the Outer Hebrides and in Shetland (Flinn 1978a and 1978b). On the basis of work by J. Geikie (1873, 1878) it has been widely accepted that the ice from the western Highlands moved west and north-west and overwhelmed the Outer Hebrides before terminating somewhere east of St. Kilda (Stewart 1933, Wager 1953). Although Geikie's interpretation was contradicted by Campbell (1873) and Bryce (1873), the mainland origin of the ice which covered the Outer Hebrides has dominated the literature for one hundred years and is even used by Denton & Hughes (1981) in their model of the last British ice sheet. Recent work by von Weymarn (1974), Coward (1977), Peacock & Ross (1978), and Flinn (1978a) indicates that during the last glaciation a local centre, or centres, of ice accumulation and dispersal developed over, or to the west of, the Outer Hebrides. Similarly, Peach & Horne (1879) suggested that Shetland was glaciated by Scandinavian ice. Hoppe (1970, 1974) disagreed with Peach and Horne in that he deduced two phases of ice movement during the last glaciation—one from Scandinavia followed by the development of a local ice cap with radial outflow. Flinn (1978b, 1978c) has suggested that Scandinavian ice crossed southern Shetland at an early stage of the last glaciation, or during a previous glaciation, but at the maximum of the last glaciation an ice mass accumulated over the Shetland Islands and flowed outwards from a north-south ice-shed.

From the above discussion it can be seen that the western and northern margins and directions of flow of the last Scottish ice sheet still remain questionable. The eastern margin is even more controversial. The flow lines shown on most maps of the last Scottish ice sheet (Figs. 3.1A, 3.1B, 3.1C, 3.7) all appear to have been deflected either north-eastwards or south-eastwards along the east coast of Scotland. These deflections have been interpreted as the result of Scottish ice meeting ice which moved westwards across the North Sea basin from Scandinavia. In estab-

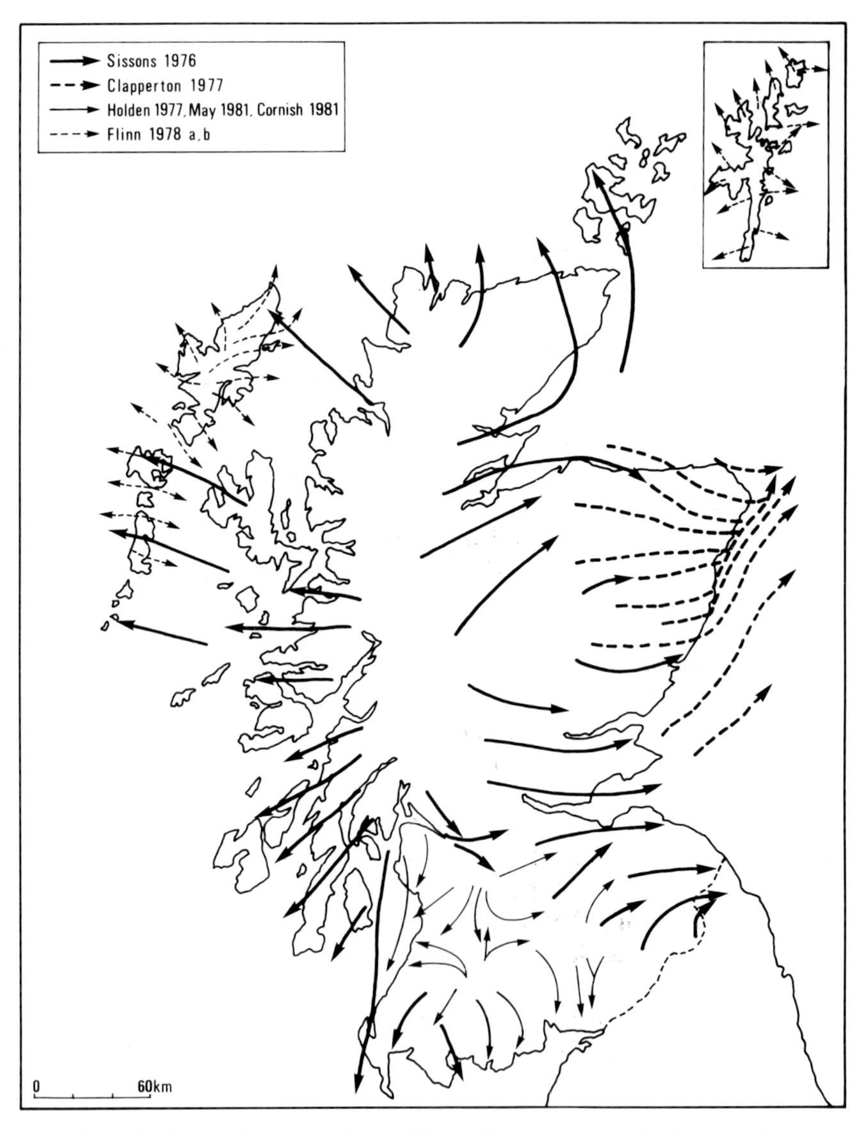

Fig. 3.7. Recent interpretations of lines of movement in the last ice sheet.

lishing the validity of this hypothesis the investigation of the deposits and morphology of the floor of the North Sea would hopefully provide some useful information. However, recent publications (Jansen 1976, Holmes 1977, Hughes *et al* 1977, Owens 1977, Thomson & Eden 1977, Eden *et al* 1978, Sissons 1981a) which refer to the Quaternary sediments of that area of the North Sea to the east of Scotland suggest two opposing schools of thought. Eden *et al* (1978, p. 2) conclude that there was complete ice cover over the central North Sea basin and that the Scottish and Scandinavian ice sheets were confluent. Sissons (1981a, p. 15) concludes, "Re-interpretation of evidence from the central North Sea indicates that the two ice sheets were not confluent during the Late Devensian." It seems, therefore, that the question of the position of the eastern margin of the Scottish ice sheet has yet to be resolved but if it was not opposed by Scandinavian ice moving westwards another cause for the deflection of the flow lines in the Scottish ice along the east coast of Scotland will have to be found.

Leaving aside the controversies about the location of centres of ice accumulation and dispersal and the position of the ice sheet margins at its maximum extent, the principal environmental implications of the development of the Late Devensian Scottish ice sheet remain unchallenged. Sometime after 30,000 bp and before 20,000 bp oceanic and atmospheric conditions in the North Atlantic underwent a major change. At that time the land area of Scotland was considerably more extensive (Fig. 3.2) and what was probably a relatively dry, cold climate changed to a colder and wetter climate so that a series of valley glaciers could develop in the high ground in the western Highlands and islands and in the western Southern Uplands. These valley glaciers expanded and thickened so that eventually (probably by about 20,000 bp) the entire mainland and most, if not all, of the islands and much of the area now covered by less than 100 m of sea water was buried beneath glacier ice. It is highly likely that the surface of this ice sheet attained a maximum altitude in excess of 1,700 m and over central Scotland the ice was probably more than 1,500 m thick. Boulton *et al* (1977) have suggested that basal temperatures of the ice sheet ranged between −6°C and −10°C and that the surface temperatures were between −20°C and −24°C. The development and wastage of this ice sheet was by far the most significant event in the evolution of Scotland's environment during the last 30,000 years. It obliterated all plant and animal life and greatly modified the land surface. Large areas experienced severe glacial erosion and were left as newly exposed bare rock surfaces after deglaciation. Large areas were covered with till and fluvioglacial sands and gravels to thicknesses in excess of 50 m. The physical presence of the ice mass resulted in considerable isostatic depression which in turn allowed, during early phases of deglaciation, the invasion by the sea of large areas which had previously been land. The remainder of this book is concerned with environmental changes in Scotland all of which are related in some way to the former presence of the last ice sheet. Even if some of these environmental changes are not directly associated with the former presence of glacier ice, they are all linked to the ice sheet because they did not become operative until the ice sheet had wasted away.

THE RETREAT AND DOWNWASTAGE OF THE ICE SHEET

Various suggestions have been made to explain the termination of the expansion of the last ice sheet. It is possible that the continued expansion of the ice sheet was stopped by a decrease in snowfall in the accumulation areas resulting from the polar front being situated well to the south of the British Isles (Sissons 1981a, p. 15). Several authors have suggested that a rapid rise in average temperature occurred about 14,000 bp concomitant with the return of subpolar waters to the west of Britain. Whatever the cause, it now seems fairly certain that once initiated the wastage of the last Scottish ice sheet was rapid. Bearing in mind the uncertainties associated with some of the earliest radiocarbon dates currently available from post ice-sheet strata in Scotland (see Chapter 4) it is likely that the entire Scottish ice sheet melted in less than 4,000 years.

Nearly one hundred years ago (A. Geikie 1887), it was realised that the dissipation of the ice sheet led to a lowering of the ice surface which in turn led to the emergence of the higher ground above the ice sheet with individual glaciers occupying the main valleys. For many years there was much debate about the relative significance of actively retreating ice fronts as opposed to downwasting ice surfaces, with the last ice occupying the valley floors and lowlands (Sissons 1958a, 1958b, Price 1960, 1963a, Price 1973 pp. 7–11). Many investigators were obsessed with determining ice marginal positions during the deglaciation of Scotland (Kendall & Bailey 1908, Clough 1910, Gregory 1915, 1926, 1927, Charlesworth 1926a, 1926b, 1955, Simpson 1933, Sissons 1961c, 1964). As recently as 1967 Sissons indicated two marginal positions associated with readvances of ice during the wastage of the last ice sheet—the Aberdeen-Lammermiur Readvance limit and the Perth Readvance limit. Sissons (1981a, p. 7) has recently stated that "Nowhere on the Scottish land area has an end moraine been interpreted as marking the limit of the last ice sheet." Similarly all the postulated readvance limits during the wastage of the ice sheet have been abandoned (Paterson 1974, Sissons 1974b). Minor readvances have been suggested at Stirling in the Forth Valley (Sissons 1974) in Loch Fyne (Sutherland in Sissons 1981a), along the coast of the north west Highlands between Ardnamurchan and the Sound of Sleat (Peacock 1970) and in Wester Ross by Robinson and Ballantyne (1979). There is no stratigraphical evidence for the Wester Ross readvance and it is based on the existence of a moraine ridge 2–8 m high which occurs intermittently over a distance of some 30 km. It is much more likely that such a moraine ridge represents a still stand of not more than a few decades rather than a major readvance.

The overwhelming evidence from the Scottish mainland points to a more or less continuously downwasting ice mass with vast volumes of meltwater being released which were responsible for the many thousands of meltwater channels cut in solid rock on valley-side slopes, and for the large quantities of fluvioglacial sands and gravels deposited on the low ground. It is also likely that much of the ice mass became stagnant in the relatively early stages of deglaciation. Once the ice sheet had thinned sufficiently for the underlying rock ridges to emerge above the ice surface, many individual ice streams would either have been completely cut off from their source areas or at least have had their accumulation areas greatly reduced in size. As the upper surfaces of individual valley glaciers were lowered to expose steep valley sides, considerable amounts of debris would fall or be washed on to the glacier's upper surface and eventually large areas of glacier ice would be buried under several metres of debris. Such a debris layer would have retarded ice wastage. Extensive areas of stagnant debris-covered ice along valley floors and along coastal lowlands may have developed their own meltwater drainage systems both englacially and subglacially. Some of these systems were controlled by extensive englacial water tables. The last phases of ice wastage were characterised by large volumes of meltwater flowing along valley floors and across the newly deglaciated lowlands. It was these waters which laid down much of the so-called valley-fill materials so characteristic of many valleys in Scotland and which were subsequently eroded by Late-glacial and Post-glacial streams to produce terraces.

It could be argued that the almost total absence of well-defined morainic ridges on the Scottish mainland stems from the great erosive power of the vast quantities of meltwater produced by the wastage of the ice sheet. Alternatively, the absence of morainic ridges may simply reflect the condition of the ice at the time when the ablating upper ice surface intersected the ground surface. If downwastage rather than frontal retreat characterised the dissipation of the ice sheet then stagnation along marginal zones would not have been conducive to moraine formation. The inability to interpret ice marginal locations over any distance and the lack of reliably dated organic material in association with any ice margins means that it is simply not possible to delimit with confidence the margin of the Scottish ice sheet at any particular stage between its maximum extent and complete disappearance.

LANDFORMS AND DEPOSITS PRODUCED BY THE ICE SHEET

Much of the evidence used to reconstruct the character of the last ice sheet to cover Scotland consists of the landforms and deposits either produced by, or at least modified by, that ice sheet. The very existence of an ice cover profoundly changes the geomorphological processes which operate (Price 1973, Sugden & John 1976). Glacier ice itself may not be a very significant erosive agent if it is clean and only moving slowly but when charged with debris and moving at rates of between 10 and 100 m per year it can be a very significant agent of erosion, transportation and deposition. During deglaciation the release of vast quantities of meltwater is also conducive to a great deal of erosion, transportation and deposition by the rapid discharge of meltwater through channels on, in, under and at the margins of the ice mass. Although the effects of each set of geomorphological processes associated with glacier ice will be examined in turn, it must be remembered that they combine to create distinctive glaciated landscapes. These landscapes consist of some landforms which were created by the ice sheet or its meltwaters along with other landforms which existed prior to the development of the last ice sheet and were only modified.

A

B

C

Glacial erosion

It is easy to envisage the accumulation of snow in valley heads and the transformation of that snow into glacier ice and the subsequent movement of the ice under the force of gravity in the form of a glacier moving down valley. The abrasive action of debris-laden ice in motion results in the modification of the valley head into a corrie and the valley into a glacial trough. The corrie and glacial trough (approximately U-shaped in cross-section) reflect the interaction between the processes of glacial erosion and the lithology, structure and morphology of the valley prior to its occupation by glacier ice. As the valley system becomes filled with glacier ice, former watersheds are overridden and ice streams may flow in directions somewhat different to those followed by the former river systems. Glacially-breached cols and through-valleys are so produced.

Much has been written about the impact of glacial erosion on the Scottish landscape, (A. Geikie 1865, 1901, J. Geikie 1894, Linton 1949, 1951a, b, 1963, Sissons 1967a, Haynes 1968, Sugden 1968, 1969). A major difficulty encountered by all these workers has been to identify the role of inherited characteristics in the present forms of glacial erosion. It has proved impossible to recognise landforms due solely to preglacial or interglacial processes and to separate the effects of the last ice sheet from those of earlier ice sheets. All the features of glacial erosion described below contain elements which may have been inherited from their preglacial form and elements created both by glacial and periglacial processes active during cold periods which preceded the Devensian Age.

Glacial troughs (Pl. 3.1A, B, C) are numerous and widespread throughout the western Highlands and Islands, the Grampians, in the Cairngorms, and to a lesser extent in the western Southern Uplands. In a sense every valley in Scotland is a glaciated valley in as much as they were all occupied by glacier ice during the last glaciation. However, in areas of high local relief which largely coincide with the principal areas of ice accumulation and dispersal, glacial erosion was most severe and valley sides were oversteepened and straightened. The control of major valley systems by the underlying geological structures also gave many of the glacial troughs strong north-east to south-west trends in the south-west Highlands and strong north-west to south-east trends in the north-west Highlands. Many of the glacial troughs extend beyond the present coastline and many contain 'basins' on their floors (Fig. 3.8). The ability of glaciers to erode enclosed basins is well-established. Deep basins (over 200 m) occur between some of the islands of the

---

Plate 3.1 (on facing page).

GLACIAL TROUGHS:

A. Loch Avon, Cairngorm Mountains.

B. Lairig Ghru, Cairngorm Mountains.

C. Glen Coe.

Inner Hebrides and in the Firth of Clyde. Similar deep basins occur in the northern part of Loch Lomond and Loch Morar. Many of these deep rock basins are associated with fault zones in the underlying bedrock. Rock basins are not confined to the western Highlands. Similar features more than 100 m deep occur near Stirling in the Forth Valley, in the Devon Valley south of the Ochil Hills and in the Kelvin Valley north-east of Glasgow.

Fig. 3.8.

The main rock basins of Scotland (Sissons 1967a).

Both at the heads and on the sides of glacial troughs large arm-chair like hollows occur as a result of the active erosion by glacier ice in valley heads. There are many hundreds of corries in Scotland (Pl. 3.2, 3.3) with steep semi-circular back walls (Haynes 1968, Sugden 1968) and sometimes rock basins occupying their floors. The steep back walls range in height from 50 m to 300 m and the corries are commonly 500 m to 1,000 m wide. The altitude of corrie floors has been used to indicate firn-line altitudes (Linton 1959, Robinson *et al* 1971, Sissons 1980) but the fact that Scottish corries have been occupied by more than one ice mass makes such studies questionable.

Plate 3.2. Corrie Toll an Lochan cut in white Cambrian quartzite overlying well-bedded Torridonian sandstone. An Teallach, Ross and Cromarty (Inst. of Geol. Sci. photograph published by permission of the Director, N.E.R.C. copyright).

Whether or not glacial troughs and corries are solely the product of the last ice sheet or represent the action of several periods of glacial, periglacial and fluvial activity does not detract from their importance in the landscape of the Scottish Highlands. In particular the west Highlands are characterised by several hundred glacial troughs which are often linked to each other by glacially breached cols (Fig. 3.9). In other areas mountain massifs are scalloped by deep corries. The Cairngorm Mountains (Fig. 3.10, Pl. 3.3) exhibit a range of glacial erosional features cut into an ancient (preglacial?) plateau surface (Sugden 1968). The difficulty of identifying

Fig. 3.9. Glacial troughs and breached watersheds west of the Great Glen.

the extent of glacial erosion achieved by the last ice sheet will remain, so long as it proves impossible to date erosional forms either absolutely or relatively.

There are large parts of Scotland where the dramatic aspects of glacial erosion are absent. There are many valleys in the Southern Uplands which, although occupied by ice during the last glaciation, can hardly be described as glacial troughs. Valleys such as Moffatdale, the Talla Valley and upper Tweedale do exhibit a degree of straightness and over steepening of valley sides not usually associated with fluvial processes. There are also large areas of the ancient foreland in north-west Scotland and in the Outer Hebrides which exhibit 'glacial roughening' which is best described by the term knob and lochan. These areas of relatively low relief clearly reflect the direction of flow of ice in the basal layers of the last ice sheet with steep rock faces pointing towards the former direction of ice flow and smooth rock surfaces facing in the opposite direction.

In the Central Lowlands, in Ayrshire, in the lower Tweed Valley and in Dumfriesshire and Galloway, there are areas in which it is difficult to separate stream-lined landforms produced by glacial erosion and those produced by glacial deposition. Many drumlinoid forms consist of glacially modified bedrock overlain by a relatively thin layer of glacial till (Menzies 1976, 1981). Although these features usually have a relative relief of less than 50 m their strong alignment and relatively steep slopes (20–30°) give large areas of Scotland a very distinctive appearance.

Glacial deposition

The most common superficial material in Scotland is glacial till. This deposit probably covers at least 60 per cent of the land area, although it is often overlain by fluvioglacial sands and gravels and peat. Till is undoubtedly the most important 'parent material' from which soils develop. Over vast areas of Scotland till has little or no topographic expression in that it simply masks the underlying solid geology. The till attains its maximum thickness on valley floors, in the Central Lowlands and on coastal lowlands. The till deposited by the last ice sheet in Scotland is usually one to thirty metres in thickness and quite frequently the 'feather edge' of a till sheet can be identified on a valley side slope by a change in gradient at the junction between till-covered bedrock and bedrock at the surface.

The till laid down by the last ice sheet is very variable in character. Its composition is largely dependent upon the rock types across which the ice responsible for its deposition passed. The majority of Scottish tills contains a variety of pebble and cobble-size particles, often subangular or angular in shape, embedded in a silty-clay matrix. In areas dominated by sandstones the till matrix is often very sandy. Some tills have been derived from fluvioglacial deposits laid down prior to the advance of individual ice lobes and therefore contain high percentages of well-rounded or sub-rounded particles.

The mechanisms involved in till deposition are numerous and range from basal release of debris by melting to flow-tills with large amounts of water associated with debris flows along ice margins (Goldthwait 1971). A discussion of the genesis of till deposits is beyond the scope of this book but their character and extent in the Scottish landscape is of considerable importance. Their most important expression is, in fact, a relatively uniform surface, either of low angle slopes on lowlands or of quite steep slopes on valley sides. In the former case they are referred to as till plains while in the latter they are often referred to simply as valley fill. The latter term, although rather vague, is very useful because on steep slopes till deposits are often re-worked, after initial deposition, by periglacial and mass movement processes to the extent that they may not retain their original depositional characteristics.

There are two groups of landforms consisting of glacial till which are usually associated with the wastage of an ice sheet. Moraine ridges usually develop along the frontal and lateral ice margins and drumlins emerge from beneath the ice having been deposited and moulded while the ice sheet was still active. It has already been pointed out that morainic ridges associated with the last Scottish ice sheet are virtually unknown (except for Wester Ross–Robinson & Ballantyne 1979). There are large areas in central and southern Scotland where morainic ridges might well be expected to occur but in fact are absent. This is either because the conditions during the wastage of the ice sheet were simply not conducive to the formation of moraine ridges or, if they were developed, they have been subsequently destroyed by fluvioglacial or periglacial processes. There is considerable evidence which suggests that the last ice sheet stagnated and downwasted over large areas and these are not conditions conducive to moraine formation.

Plate 3.3. Vertical aerial photograph (Department of Geography, Glasgow University) of the corries and troughs cut into the Cairngorm Mountains.

A. Coire an Lochain
B. Coire an t-Sneachda
C. Coire Cas
D. Lairig Ghru
E. Loch Avon
F. Loch Etchachan

Drumlins (Pl. 3.4) are very common features in central and south-west Scotland and in the lower Tweed Valley. They consist of elongated mounds 10–30 m high and between 200 and 1,000 m in length. They often have steep stoss ends facing the former direction of ice movement and gentle streamlined tails (Fig. 3.11). They are often not simple features; smaller drumlins having been superimposed on larger ones (Rose & Letzer 1977). Some drumlins consist entirely of glacial till while others have a bedrock core or consist partly of fluvioglacial sand and gravel

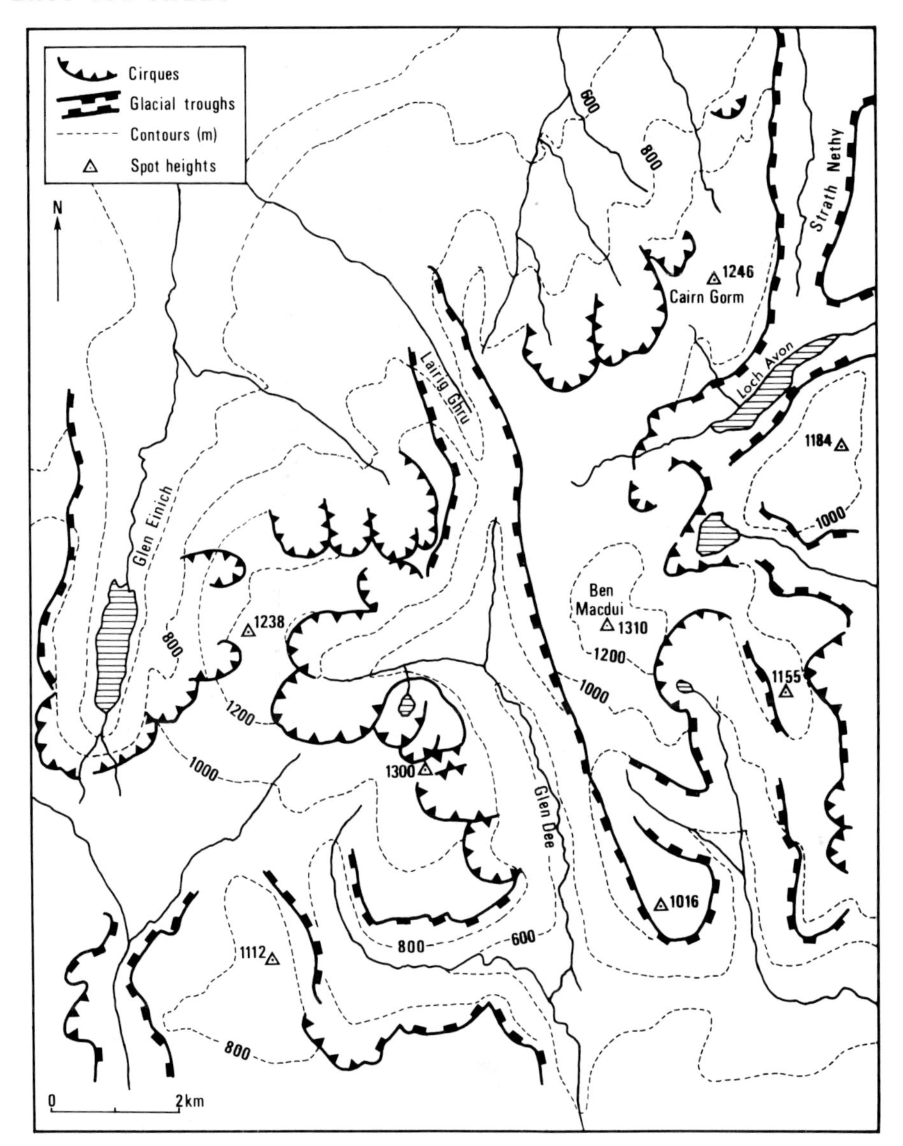

Fig. 3.10. Glacial cirques and troughs in the Western Cairngorms.

(Menzies 1976, 1981). Because of their streamlined form drumlins are extremely good indicators of the former direction of flow of the glacier ice which produced them. They also produce a very distinctive landscape with a strong lineation, local relief of 10–30 m and quite steep slopes of 15°–30°.

Although glacial till is an extremely important parent material for soil development in Scotland, detailed studies of its lithology and structure are rare. There have been numerous studies of till fabric (Kirby 1968, Young 1969, Menzies 1976) and the source areas of the coarse fraction in till have long been studied as indicators of ice movement. Quite large areas in Caithness, the south shore of the Moray Firth, Ayrshire and the Rhinns of Galloway are covered by till which contains marine shells indicating that the glacier ice had crossed areas formerly covered by the sea before depositing the till. Shells have been found in tills in the Muirkirk area, Ayrshire, at altitudes in excess of 300 m (Smith 1898, 1902).

Glacial till deposited by the last ice sheet is not confined to the present land area of Scotland. Investigations in the Minch (Binns *et al* 1973) and the central

Plate 3.4.

Drumlin near Glasgow.

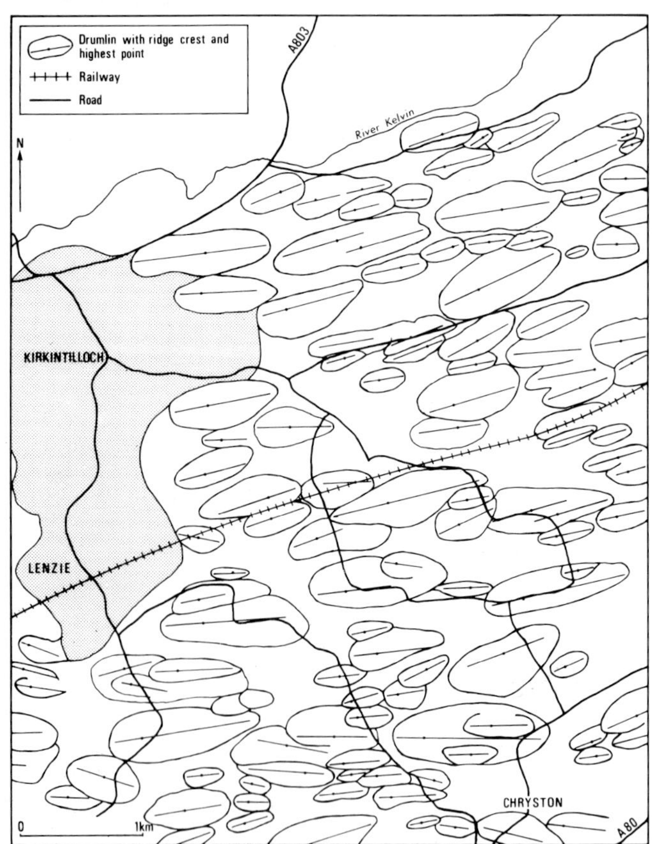

Fig. 3.11.

Part of the central Scotland drumlin field. (Based on unpublished field map by J. Rose.)

North Sea (Holmes 1977, Owens 1977, Eden *et al* 1978) reveal the depositional activity of the ice sheet in those areas.

Although not a spectacular feature of the environment, the extensive cover of glacial till produced by the last Scottish ice sheet has played a very important part in the evolution of the Scottish environment. Not only has soil development been greatly affected by the presence or absence of till but the nature of surface and subsurface drainage, the stability of slopes and the detailed surface morphology of large areas have all been influenced by the character and thickness of the till sheet. Although the features of glacial erosion are far more dramatic elements in the Scottish landscape and the locations at which that eroded material was finally deposited by the glaciers may not be so spectacular in terms of their landforms, they may be important elements in the evolution of soils, vegetation and land use patterns.

Fluvioglacial erosion

The wastage of the last ice sheet produced vast volumes of meltwater which were concentrated into channels cut either on, in or under the ice, at the ice rock interface or in the proglacial areas. Until twenty years ago the significance of the erosive power of these meltwaters either as a means of creating landforms or as evidence of the former extent of glacier ice and its character was only poorly understood in Scotland. A series of papers by Sissons (1958a, 1958b, 1960, 1961a, 1961b) introduced ideas developed in Scandinavia by Mannerfelt (1945, 1949), Hoppe (1950, 1957), Gillberg (1956) and Gjessing (1960), to the interpretation of landforms produced during deglaciation in Scotland. Detailed mapping has revealed that many thousands of meltwater channels were cut by the meltwaters of the last ice sheet all over Scotland (Price 1960a, 1963a, Clapperton 1968, 1971, Young 1974, 1975, 1977).

Waters overflowing from ice-dammed lakes occurring along the lateral margins of valley glaciers have long been suggested as the cause of fluvial channels unrelated to any normal drainage pattern. As early as 1863, Sir Archibald Geikie (1863, p. 28), suggested that some channels in the Tweed Valley near Drumelzier, and others in the Manor Valley, Peeblesshire, ". . .seem as if they have been formed when the lateral glens were dammed up so as to form lakes, and the pent-up waters escaped by cutting out channels for themselves by which they escaped into the main river." J. Geikie (1869, p. 18) writing about the same channels suggested that, ". . .they should be referred to a time when the drainage of the district was greatly modified by large accumulations of snow and ice." Many of the early workers in Scotland recognised anomalous channels and channel systems in locations on valley sides, cutting through spur crests and the floors of high level cols, which are quite unrelated to the present drainage pattern. Many of these channels contain no streams while some contain streams which are unrelated in size to the channels they occupy. So long as the explanation of meltwater channels relied heavily on their association with former ice-dammed lakes, then their significance in the interpretation of former positions of ice margins and the mode of deglaciation remained poorly understood. The evidence for the former existence of ice dammed lakes in association with many meltwater channels simply does not exist. Proven lake-floor deposits, shoreline features and deltas are extremely rare. Channels with up-down long profiles, the absence of channels in localities where they would be expected to occur on the basis of the lake hypothesis, and the occurrence of channels on the reverse slopes and minor summits of spurs are very difficult to explain if lake overflows are the only means of producing meltwater channels. It is much more likely that both individual channels and channel systems are the product of meltwater streams which develop on, in, under and at the lateral margins of downwasting glacier ice (Fig. 3.12).

Meltwater channels in Scotland have been cut both in solid rock and in glacial and fluvioglacial deposits. Many channels cut in rock are 5–10 m deep, 10–30 m wide and a few hundred metres long. There are some very large meltwater channels cut in rock to depths in excess of 50 m and which are several kilometres in length (Sissons 1958a, 1963). Characteristics that make meltwater channels distinctive from other fluvial channels are as follows:

1. They begin and end in areas that under normal fluvial activity would not be likely locations for the initiation and termination of channelled flow.
2. They frequently cut across present drainage divides.
3. They sometimes have open in-take ends that bear no similarities to other types of water-eroded channels.
4. They frequently have very steep sides and are either unusually straight over considerable distances or develop tight meanders over short distances.
5. They do not usually widen appreciably in the down stream direction.
6. Tributary channels are often ungraded at their junction with the main channels.
7. They either lack a present stream compatible with the size of the channel or they are completely dry.
8. Shoreline remnants, lake-floor deposits, deltas, kame terraces and eskers can occur at either channel in-takes or terminations.

Individual channels are often difficult to recognise but when numerous channels form part of a system (Fig. 3.13 and Pl. 3.5) then greater certainty can be attached to their interpretation. It must be remembered that during deglaciation a very complex meltwater drainage system would have developed both on, in, under and at the margins of the ice mass. Only where the meltwater streams came in contact with the sub-ice surface would meltwater channels be cut in the drift or solid rock. The channels which can be observed only represent a part of the meltwater drainage system, much of which was on or in the glacier ice itself. In some situations fluvioglacial deposition replaced fluvioglacial erosion and channel segments are linked by ridges of sand and gravel (eskers) being the depositional casts of meltwater streams flowing in tunnels in the ice.

The position of many meltwater channels was often determined by the location and character of the lateral ice margin, by the lowering of supraglacial and englacial streams to the underlying topography and by the presence of large englacial and subglacial streams flowing under hydrostatic pressure. Detailed observations in the

Tweed Valley (Price 1960, 1963a, b) and in the Cheviot Hills (Clapperton 1968, 1971) have shown the importance of the superimposition of a meltwater drainage system, developed on and in the ice, on to the underlying topography.

The initial direction of meltwater drainage was determined by the slope of the ice surface but as this was lowered by ice wastage the underlying ridges, spurs and cols began to interrupt the regional flow (Fig. 3.14). Frequently it is possible to recognise the earliest direction of meltwater flow, and therefore the slope of the ice surface, by the orientation of the highest channels in a given area. These high level channels also indicate the minimum altitude of the ice surface in a particular area and the general direction of ice movement. However, as deglaciation progresses the control of the underlying topography becomes greater and there is a tendency for meltwater flow to be concentrated at ice marginal and submarginal positions along valley sides. A good example of this topographic control of meltwater movement can

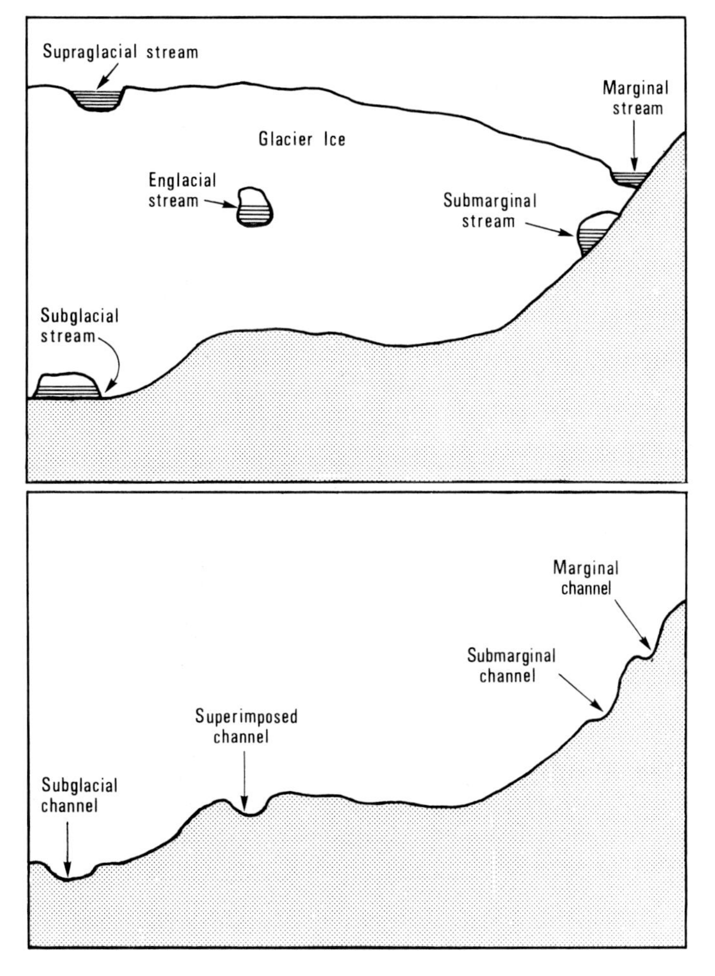

Fig. 3.12.

Meltwater streams and the channels which remain after deglaciation.

Plate 3.5A

Plate 3.5B

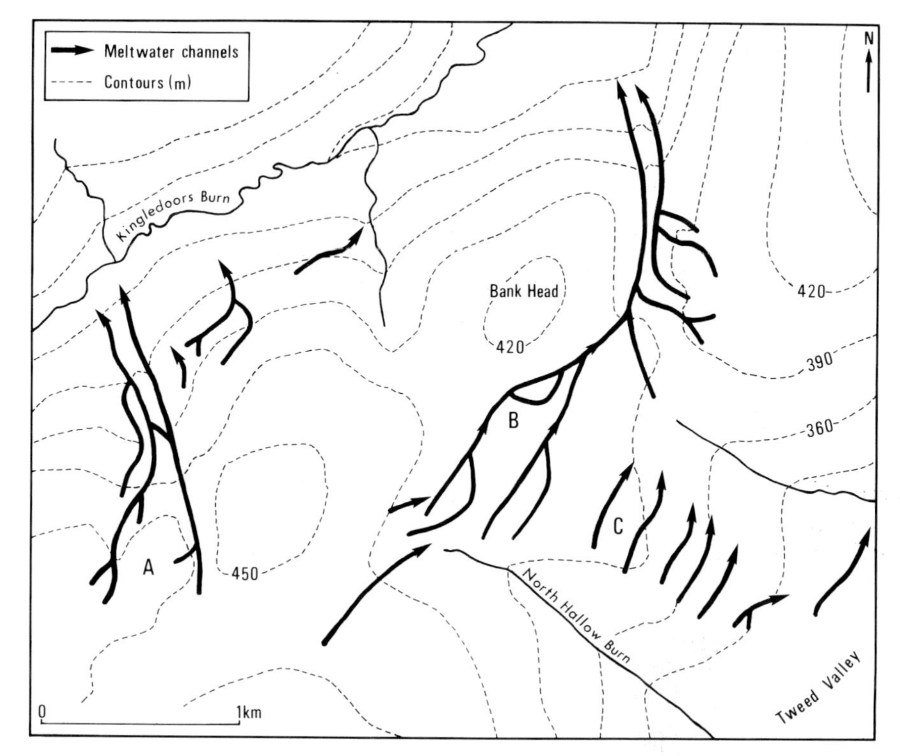

Fig. 3.13. Meltwater channels in the Upper Tweed Valley.

be seen on the south side of the Campsie Hills (Fig. 3.15). At the higher elevations (i.e. in excess of 250 m) the trend of the channels is roughly west-east and individual channels are quite long (1–3 km) and of gentle gradient, but at lower altitudes individual channels are quite short (100–500 m) and usually turn sharply down slope.

The identification of the position of formation, in relationship to the ice margin, of individual channels is often impossible except for those which are at right angles to contour lines and are therefore certainly subglacial. It has been suggested that

---

Plate 3.5 (on facing page)

A. Vertical aerial photograph (R.A.F. photography, Scottish Development Department) of meltwater channels in the Upper Tweed Valley (see Fig. 3.13).

B. View along channel B (Fig. 3.13) towards the south-west.

meltwater channels which closely follow contour lines were formed at the ice margin. However,.such channels could have been formed in a submarginal (i.e. subglacial) position.

The concentration of meltwater movement through pre-existing cols is not surprising (Clapperton 1968) but the superimposition of a meltwater stream from a channel or tunnel developed in glacier ice across the top of a spur crest (Price 1960) consisting of solid rock seems less likely. However, many hundreds of meltwater channels have now been mapped in Scotland which begin on the 'upglacier' side of a spur, cut through the spur crest and continue down or along the lee side of the spur. There can be little doubt that a complete drainage system developed on and in glacier ice can be superimposed on to the subglacial topography. In some cases both the erosional and depositional phases of such a drainage system are preserved in the landscape after deglaciation. Young (1974) has described and mapped (Fig. 3.16) a very complicated system of meltwater channels and eskers along the north-west face of the Cairngorm Mountains near Loch Morlich (Pl. 3.6). He identifies three phases of ice wastage: (a) gradual thinning; (b) widespread wastage and stagnation; (c) total stagnation. In the early phases, ice controlled drainage routes were dominant and

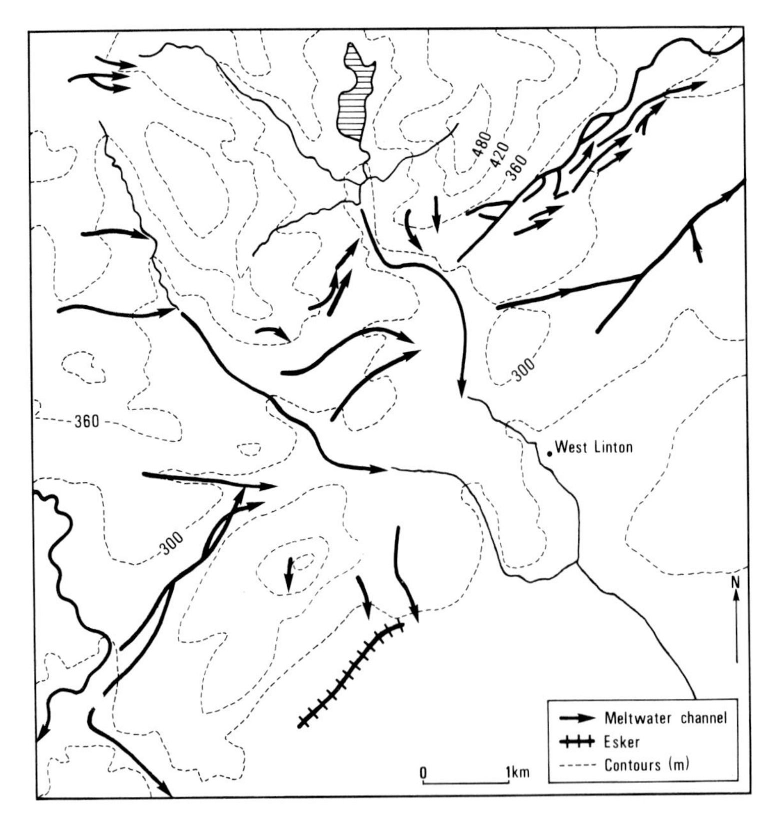

Fig. 3.14. A superimposed meltwater channel system near West Linton, Peeblesshire.

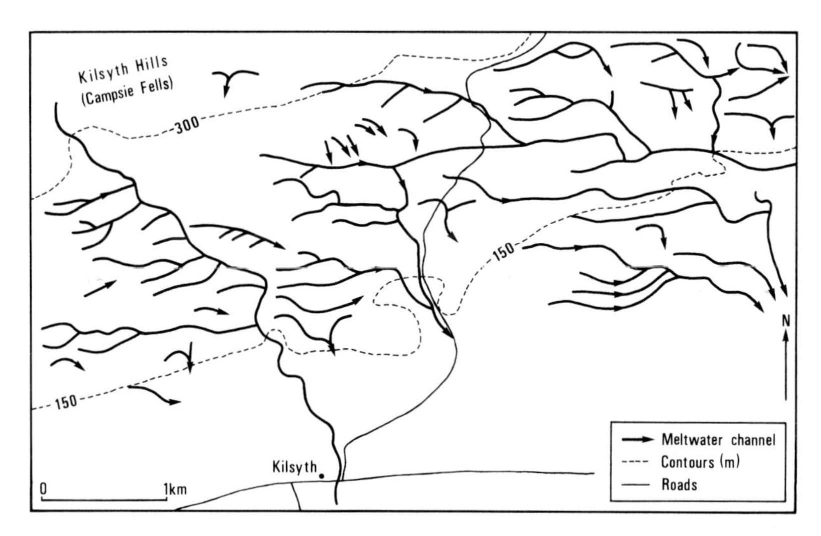

Fig. 3.15. Meltwater channels on the south side of the Campsie Fells (Kilsyth Hills).

marginal and submarginal channels were cut. Young believes that meltwater penetration of the ice mass was of the order of 160–165 m at this stage. During later phases, subglacial routes were strongly developed and depositional activity at the distal ends of channel systems is clearly represented by esker and kame complexes to the west of Loch Morlich. There is often an altitudinal zonation of features produced by fluvioglacial erosion and fluvioglacial deposition in a particular area. Meltwater channels terminate at a particular altitude and only below that altitude does fluvioglacial deposition occur. Such a relationship may well be produced by the development of an englacial water-table within the wasting glacier ice. The altitude of this englacial water-table may be controlled by the altitude of a col (Sissons 1958b), or temperature or structural conditions within the ice.

From the above discussion it can be seen that the detailed description and mapping of meltwater channels can reveal a great deal about the former extent and condition of the last ice sheet during deglaciation. Although meltwater channels are not usually spectacular features in the landscape their widespread occurrence all over Scotland has proved to be of considerable value in interpreting both the former extent of the ice sheet, its regional gradient and its condition during wastage. The importance of the lowering of the ice surface to allow ridges and spurs to become ice free has been widely established. It is likely that large parts of the ice sheet became stagnant during early phases of deglaciation and that englacial and subglacial meltwater drainage became the main means of meltwater discharge. Such a model of deglaciation at least partly explains why clearly identifiable morainic ridges are very rare in Scotland. It also explains why it is so difficult to determine the location and chronology of ice marginal positions during deglaciation. It may well be that some of the last ice to melt was located on the floors of deep and sheltered valleys with delay in final melting being assisted by a protective cover of glacial and fluvioglacial debris.

Meltwater channels may be very common features in Scotland but because they have dimensions measured only in tens or hundreds of metres they are not dramatic features in the landscape. Where they occur as interconnected systems they produce a corrugated surface with local relief values ranging between 10 and 30 m. They have often been used as route-ways by road builders and railway engineers. They make excellent sites for small reservoirs and they have sometimes been occupied by post-glacial river systems. The relationship of major valleys to the meltwater discharged by the last ice sheet is often difficult to determine. There can be little doubt that the meltwaters which cut the thousands of channels on valley side slopes eventually made their way along the floors of the major trunk valleys. The impact of these meltwaters on the morphology and channel geometry of the

Plate 3.6. Vertical aerial photograph of meltwater channels on the northern slopes of the Cairngorms between the Lairig Ghru and Loch Morlich (Department of Geography, Glasgow University).

major Scottish river systems is poorly understood. In very general terms it is usually suggested that the main effects produced by the discharge of large volumes of meltwater and glacial debris through the river system was a net accumulation of alluvium on the valley floors which has been steadily eroded during postglacial time to produce river terraces.

The implications of the existence of many thousands of kilometres of channels cut by the meltwaters of the last ice sheet have only recently been appreciated. Unfortunately there is no reliable means of determining exactly when all this meltwater erosion took place. Any individual channel may have been cut in one ablation season (three or four months) and then abandoned. Other channels may have been cut during previous glaciations and re-occupied. It would even be helpful to know if the main meltwater activity took place over a period of 4,000 years or perhaps was much more dramatic and lasted only 1,000 years.

Fluvioglacial deposition

In a glacierised environment a great deal of rock debris is incorporated into the ice mass and is lodged at the ice-rock interface. During deglaciation this debris is released and is added to by slumping and sliding of material across valley side

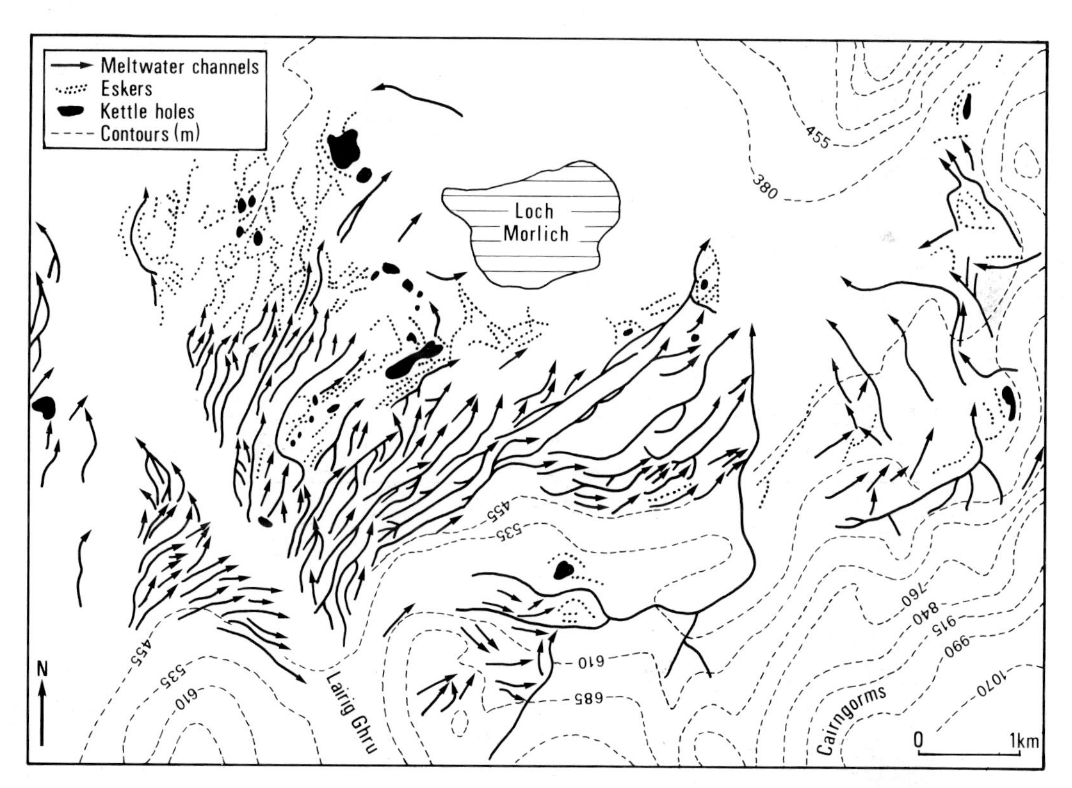

Fig. 3.16. Fluvioglacial features south of Loch Morlich, Spey Valley (Young 1974).

slopes and on to the wasting ice as the ice surface is lowered. In a situation in which valley floors remain occupied by ice when ridge crests and spurs become ice free there is a great tendency for meltwater streams to collect large amounts of debris and to transport it along the lateral margins, across the ice surface and through tunnels in and under the ice towards the proglacial area. The meltwater drainage system associated with a downwasting and probably largely stagnant ice mass is capable of transporting and depositing large quantities of fluvioglacial material. Such drainage systems are characterised by large seasonal and daily fluctuations in discharge linked to fluctuations in ablation determined to a large extent by seasonal and daily temperature changes. There are many similarities betwen drainage systems developed on wasting ice masses and those which occur in semi-arid climates. In both environments flood discharges are responsible for the movement and deposition of large quantities of material.

The character of fluvioglacial deposits is largely determined by the nature of the parent material (largely glacial till or periglacial slope deposits), the distance of transport and the environment of deposition. Since all fluvioglacial deposits are the product of transportation by and deposition in water, the rock particles are rounded or sub rounded in shape. The material ranges in size from large boulders and cobbles to fine sand, silt and clay. Channel deposits consisting of gravel and sand with cross bedding are extremely common while in some areas, although not common in Scotland, vast amounts of material of the silt and clay size accumulate in temporary ice-dammed lakes. A great deal of fluvioglacial material arrived at the coast of Scotland and is now found incorporated in beaches or deposited in the off-shore zone.

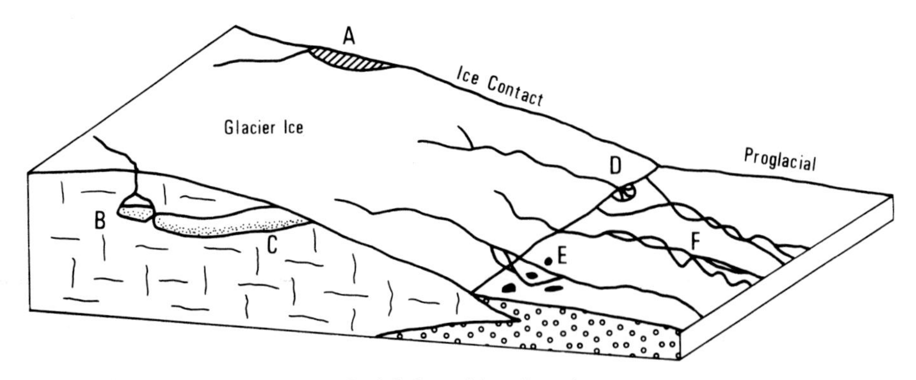

Fig. 3.17. Ice-contact and proglacial depositional environments.

Fluvioglacial deposits are usually classified into two principal groups (Fig. 3.17) on the basis of whether they were laid down in contact with the glacier ice (ice-contact deposits) or beyond the ice margin (proglacial deposits). Ice-contact deposits accumulated in channels on the ice surface or in tunnels in or under the ice (Price 1973, pp. 131–171). These deposits can either be the product of flowing water confined between ice walls and simply let down onto the underlying topography as the ice wasted away, or they are produced by meltwater streams carrying material either into caverns within or under the ice or into lakes at the ice front or even on the ice surface. The morphology of deposits laid down in contact with the ice is therefore determined not only by the original shape of the accumulation but by the subsequent wastage of the ice upon or against which the sediment accumulated (Fig. 3.18).

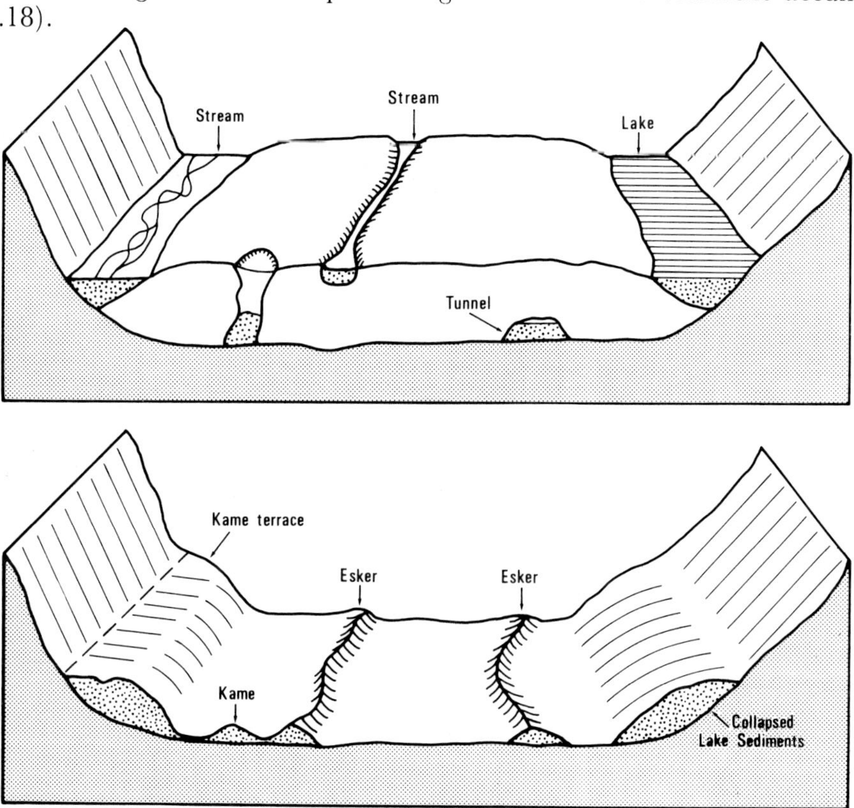

Fig. 3.18. The formation of kame terraces, kames and eskers.

Ridges of sand and gravel representing the former courses of meltwater streams on, in or under the ice are known as eskers (Price 1966, 1969). They are commonly 5–30 m high and consist of well-rounded gravel and sand with depositional structures characteristic of channel deposits. They often exhibit tributary junctions and meander bends and have irregular crest lines (Pl. 3.7). Observations of eskers in the process of formation in areas currently undergoing deglacierisation have established that eskers which originate by deposition of material in channels on, or in, an ice mass can survive being let down on to the subglacial surface during deglaciation without having their ridge form destroyed.

Kames consist of fluvioglacial material which accumulated in small lakes on the ice surface or in caverns in or under the ice. They can occur as simple conical mounds in isolation or as groups of mounds of varying shapes and sizes. There is often a gradation between a group of kames and what is known as kame and kettle topography or pitted-outwash (Fig. 3.19). Along a retreating ice margin the

Plate 3.7. Esker ridges near Loch Walton, Campsie Fells.

concentration of meltwater flow in supraglacial and englacial channels can produce an outwash fan which can extend up onto the ice surface (Price 1970, 1971). The presence of glacier ice either in the form of a sheet or a series of detached ice blocks beneath the proximal part of an outwash fan can result in the development of depressions (kettle holes) in the surface of the fan when the underlying ice melts out. Another ice marginal location which produces a distinctive landform resulting from fluvioglacial deposition is the junction between a valley side and a downwasting ice mass. Lateral meltwater streams can accumulate quantities of sand and gravel which partly rest on the hillside and partly on the ice surface. When the ice finally wastes away terraces, sometimes containing kettle holes, are left perched on the valley side (Fig. 3.18).

Proglacial fluvioglacial deposition is usually much more extensive than ice-contact deposition. Enormous quantities of sand, gravel, silt and clay are carried beyond the retreating ice margins. Previously deposited glacial and ice-contact deposits are re-worked by the proglacial streams to create valley sandar (valley trains or valley outwash spreads). The proglacial streams often destroy moraine ridges, eskers and kame and kettle topography, particularly in valley bottoms and lowland areas, so that only a small percentage of the depositional landforms produced during the earlier phases of glaciation survive the final onslaught of the meltwaters. The outwash spreads themselves are eventually dissected as the volume of debris being supplied to the drainage system begins to decrease.

In parts of North America where the general topography allowed the development of large proglacial lakes great quantities of fluvioglacial material were deposited as bottom deposits in these lakes. The subsequent drainage of these lakes produced extensive flat plains underlain by silts and clays. No really extensive areas of proglacial lake development associated with the retreat of the last ice sheet have been identified in Scotland. Almost certainly, ice-dammed lakes would have existed within the Highlands and Southern Uplands during the period of ice sheet wastage, similar to those that have been identified during the Loch Lomond Stadial (see Chapter 4). As early as 1877, Bell suggested the existence of an ice-dammed lake in the lower Clyde Valley and more recently Brown (1980a) has made reference to a 'Clydesdale' lake. However, widespread lake-floor deposits and lake shorelines are not common features in Scotland and it is most likely that large quantities of silt and clay were deposited in the off-shore environment.

Undoubtedly the most extensive surface expression of fluvioglacial deposition associated with the wastage of the last ice sheet in Scotland is simply a uniform and relatively featureless spread of sand and gravel. These gravel spreads range in thickness from one metre to fifty metres. Their upper surfaces are now often masked by a layer of peat. Although relatively featureless at present, immediately after their abandonment by the meltwater streams which produced them their surfaces would have had local relief values of one to ten metres in the form of abandoned channel systems or kettle holes. The post-glacial weathering and mass-movement processses and subsequent soil and peat development have obliterated many of these irregularities. In some cases, outwash spreads can be traced up-valley where they merge into kame and kettle complexes (pitted outwash) and down-valley where they overlie raised marine sediments. Such relationships were described by J. B. Simpson (1933) in the lower Earn and Tay Valleys and by Sissons (1963b) in the Forth Valley. Along the edge of the south-eastern Grampians a series of outwash fans (sandar) are to be found where the highland valleys open out into Strathmore. Near Blairgowrie, for example, an outwash fan covers some 25 sq km and contains kettle holes in its proximal part which consists mainly of gravel and has a gentle gradient towards its distal part which is underlain mainly by sand.

In central Scotland, the Forth Valley, the Kelvin Valley, the lower Clyde Valley and much of the Lothians and Fife are underlain by fluvioglacial gravels either laid down during the advance or retreat of the last ice sheet. It must be remembered that fluvioglacial deposition takes place both during the expansion and dissipation of an ice sheet. In the Kelvin Valley, north-east of Glasgow, for example, up to 60 m of sands and gravels are overlain by the till deposited by the last ice sheet. The till is generally one to three metres thick and it in turn is overlain by up to 15 m of sands and gravels in the form of outwash fans or kames and eskers.

The terms 'kame' and 'esker' have for long been used interchangeably in Scotland. In Scottish place names the term 'kame' has often been used in association with an elongated steep-sided mound. For this reason the major ridges of sand and gravel in the vicinity of Carstairs, Lanarkshire (Gregory 19155a, Ross 1925, Charlesworth 1926, McGregor 1927, Goodlet 1964, McLellan 1969) and between

Falkirk and Linlithgow in the vicinity of Polmont (Gregory 1913) have traditionally been referred to as kames. They are in fact classic esker systems. The Carstairs Kames consist of a series of ridges up to 24 m high and some 6 km in length aligned in a west-south-west to east-north-east direction (Fig. 3.20). Numerous exposures associated with commercial gravel extraction indicate that the deposits are entirely water-laid with the waters moving from a south-westerly source. Sissons (1961) and McLellan (1969) have both concluded that these ridges are of subglacial origin and are esker ridges representing a part of a very extensive englacial and subglacial drainage system in Southern Upland ice.

Other large esker ridges occur on the Moray Firth lowlands between Croy and Nairn, near Dornoch and between Golspie and Brora. Much more common throughout Scotland are complex esker and kame and kettle systems with individual esker ridges attaining heights of 5–15 m and lengths of a few hundred metres. Such features have been described in detail in the Spey Valley (Young 1974, 1975, 1977) and occur widely in central Scotland (e.g. the south face of the Campsie Hills—Fig. 3.15—the Eddleston Valley, Sissons 1958). It is often difficult to separate small esker ridges from kame and kettle complexes but together they indicate the former presence of stagnant ice honeycombed with passages and caverns in which fluvioglacial deposition took place.

The development of extensive areas of fluvioglacial deposits during the wastage

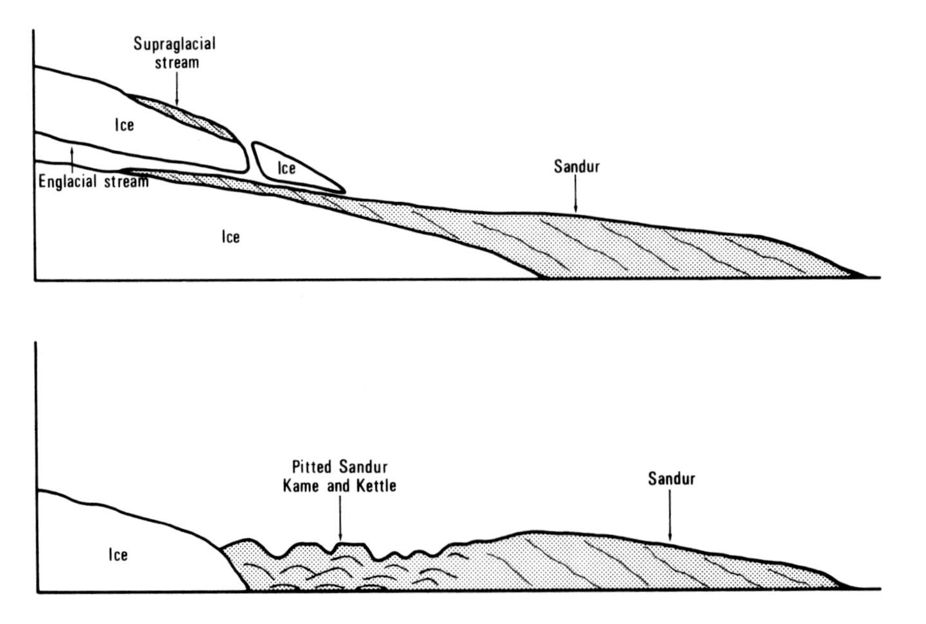

Fig. 3.19. The formation of a pitted-sandur (kame and kettle topography) as a result of the melting of buried ice.

of the last Scottish ice sheet made important contributions to the character of the Scottish environment in several ways. In morphological terms the meltwater streams created three types of landforms by their depositional activities. Considerable areas of low angle slopes underlain by well-drained sands and gravels are associated with outwash spreads. Over large areas in central Scotland and on many wide straths in the Highlands, these deposits represent the only viable intensive agricultural land. These same fluvioglacial deposits have proved to be excellent sources of aggregate for the construction industry (McLellan 1967a, b, McLellan 1970, Chester 1980) and attractive sites for settlement and urban expansion. Secondly, the depositional casts of major englacial and subglacial streams have been left in the landscape as distinctive sinuous, steep-sided well-drained ridges. These eskers frequently are the only well-drained ground and their distinctive vegetation cover makes them conspicuous elements in the landscape. Because esker ridges contain relatively high percentages of rounded gravel they have proved particularly attractive to the aggregate industry and many large eskers have been partially or completely destroyed during the last twenty years. Thirdly, there are extensive areas, particularly on valley side slopes, covered by a chaotic assemblage of ridges and mounds consisting of sand and gravel. Kames, kettle holes and short esker ridges produce a very irregular surface with steep slopes (20°–30°) surrounding elongated and circular peat-filled depressions. Such areas have often been confused

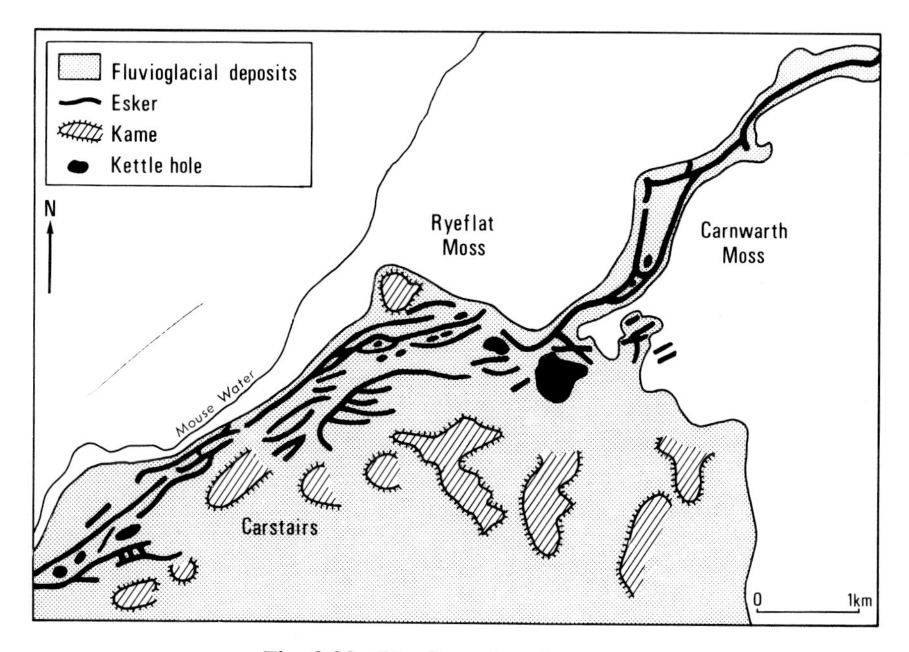

Fig. 3.20. The Carstairs esker system.

in the past with morainic accumulations and many of the so-called ice margins and re-advance limits, which earlier workers believed they identified, were in fact based on the location of these fluvioglacial complexes.

Much of the rock debris carried by meltwater streams was deposited in what is now the off-shore zone. Certainly much of the gravel and sand both in the raised beaches and present beaches of the Scottish coastline represent the re-working of fluvioglacial and to a lesser extent glacial deposits. Estuaries such as the Tay, Forth and Clyde, the channels between the inner Hebridean islands and the Minch all have large quantities of fluvioglacial sand and gravel on their floors. Detailed investigations by the Institute of Geological Sciences have revealed the extent and character of these deposits (Binns *et al* 1974, Deegan *et al* 1973, Peacock 1974, Peacock *et al* 1980).

SEA-LEVELS

Information about the relationship between the retreating margins of the last ice sheet and the sea is very limited. At the maximum stage of the ice sheet world sea-level would be relatively low but Scotland would have been isostatically depressed owing to the presence of the ice cover. The extent of that isostatic depression is unknown but could have been of the order of hundreds of metres (Denton & Hughes 1981). Whether or not the outer margins of the last Scottish ice sheet terminated on the dry continental shelf or were marine based has not yet been established. Similarly, it is not known if, once retreat began, the relationship between isostatic rebound and the eustatic rise in sea-level was such that the sea invaded the outer parts of the continental shelf as the ice retreated. Information about the chronology of the changing position of the Scottish coastline during the wastage of the ice sheet is very sparse and controversial (Sissons 1981, pp. 1 and 2). Radiocarbon dates from a bore hole on the sea-bed near Colonsay (Harkness and Wilson 1974) and from the bed of the North Sea some 80 km east of Aberdeen (Holmes 1977) suggesting that these areas were ice free by 16,000 and 17,000 bp respectively have to be discounted because of likely contamination of the dated material. At present therefore, little can be said about the retreat of the ice margins and associated sea-levels until the ice margins were located in the general vicinity of the present coastline, probably sometime after 14,000 bp (see Chapter IV).

CLIMATE DURING DEGLACIATION

There is no direct evidence of climatic conditions in Scotland between 18,000 and 14,000 bp. The geomorphological evidence does suggest that there was rapid ice wastage which produced large volumes of meltwater in association with large areas of downwasting and probably stagnant ice. The wastage and stagnation of the ice sheet could have been caused by a decrease in snowfall or an increase in temperatures or a combination of both. Recent work on ocean/ice/atmosphere interactions in the North Atlantic during periods of deglaciation (Ruddiman & McIntyre

1981a, b), based on 39 cores, 48 radiocarbon dates and 28 ash zones for stratigraphic control, allow paleoenvironmental reconstructions over the period 20,000–10,000 bp. From 20,000–16,000 bp cold oceanic waters extended south to the polar front at 45°N. Foraminiferal productivity in this area was very low in the interval 16,000 to 13,000 bp but the productivity abruptly rose at about 13,000 bp as the polar front retreated northwards. Ruddiman & McIntyre (1981, p. 128) state, "We infer that the interval 16,000 to 13,000 yr B.P. marked the most rapid delivery of meltwater and icebergs to the subpolar North Atlantic . . . . We estimate that more than half of the total deglacial influx of meltwater and icebergs entered the North Atlantic between 16,000 and 13,000 yr B.P." There is considerable evidence pointing to high summer insolation values after 17,000 bp in the northern hemisphere (Milankovitch 1941, Berger 1978) and to relative dryness over much of north-west Europe at least until 13,000 bp. However Ruddiman and McIntyre point out that (1981, p. 129), "solar insolation alone is even less likely to be a viable explanation of the speed and timing of early deglaciation." It has been suggested by various authors (Hollin 1962, Hughes *et al* 1977, Andrews & Barry 1978, Denton & Hughes 1981) that marine attack on vulnerable portions of a continental ice sheet is a very significant mechanism in deglaciation. As a result of rising sea-level, stabilising ice shelves around the periphery of an ice sheet are removed and increased calving of marine glacier terminii leads to increased rates of ice movement in the ice streams leading to those terminii. The net result is the general thinning of the interior of an ice sheet by means of the removal of ice by calving icebergs into the ocean. Evidence from the sediment cores in the North Atlantic supports the suggestion of increased meltwater and iceberg influx over the period 16,000–13,000 bp. The surface layers of the North Atlantic ocean were chilled and reduced in salinity by the influx of icebergs. Such a situation would allow the development of extensive sea ice in winter down to latitude 45°N (Ruddiman & McIntyre 1981) along with a more southerly alignment of the major storm tracks and a low level of moisture delivery to north-west Europe.

From the above evidence it can be deduced that Scotland probably experienced cold dry winters and warm dry summers during the period 16,000–14,000 bp. Rapid deglacierisation occurred because of a combination of decreasing snow fall, increased insolation and increased rates of movement of those glaciers which terminated in the rising sea leading to thinning and ultimate stagnation of the interior of the ice sheet. At the same time relative sea-level was rising and extensive calving at the ice sheet margins took place. It was likely that sea-ice developed during the winters and that oceanic temperatures to the west of Scotland remained low and therefore only limited moisture was brought to Scotland from that source. Throughout the period 18,000–14,000 bp the major storm tracks across the North Atlantic were located well to the south of the British Isles. Sometime prior to 13,000 bp, by which time the polar front had retreated to the area south-west of Iceland, the oceanic and atmospheric conditions to the west of Scotland underwent a major change which led to rapid increases in temperature which led to the final dissolution of the ice sheet and the beginning of the Lateglacial Interstadial. Until such time as either more information about the location of the polar front between 20,000 and 13,000 bp is obtained and sea surface temperatures are known for that period or, the chronology of the re-occupation of what are now the off-shore areas of Scotland by the rising sea-level associated with ice sheet wastage is known, it will not be possible to determine the character of the Scottish climate during deglaciation in greater detail.

CONCLUSIONS

The most important event in the evolution of Scotland's environment over the last 30,000 years was the development of the ice sheet which completely buried both the mainland and surrounding island groups by about 20,000 bp. This ice sheet probably took less than 10,000 years (possibly only 5,000 years) to develop.

The ice sheet was probably initiated as a result of lowering of the snowline concomitant with the migration of the oceanic polar front and associated storm tracks southwards to latitudes to the south of the British Isles. It is likely that the zone of heaviest snow fall and of most rapid build up of glacier ice occurred first in the western Highlands and migrated southwards to the Southern Uplands. The centres of ice dispersal in the western Highlands were dominant to the extent that ice from those sources covered all of the Highlands and much of central Scotland. Secondary centres of ice accumulation developed slightly later in the central and western Southern Uplands and in the Outer Hebrides.

The impact of the ice sheet on the geomorphology of Scotland was considerable. Landforms produced by fluvial activity in the preglacial and interglacial periods and by glacial and periglacial processes in an unknown number of previous glacial periods were modified by glacial erosion. The sharp topography of glacial troughs, breached watersheds and corries so characteristic of much of the Scottish Highlands and to a lesser extent of the central Southern Uplands is a product of erosion not only by the last ice sheet but also of earlier periods of glaciation. Although less dramatic, the deposits laid down by glacier ice and meltwaters cover a large percentage of Scotland and constitute important constructional landforms and the parent material for soil development.

It is likely that once ice sheet wastage began it progressed rapidly and was assisted, at least in its earliest phases, by a rapid rise in relative sea-level. Although reliable dates for deglaciation of land areas around the periphery of the Scottish ice sheet are not available, it seems likely that the ice sheet had thinned and retreated considerably from its maximum extent by 14,000 bp.

The complete destruction of plant and animal life in Scotland by the last ice sheet now seems certain. Lack of dated organic deposits earlier than 14,000 bp make impossible any detailed interpretations of the stages of deglaciation and their associated environmental changes.

4

THE LATEGLACIAL PERIOD
circa 14,000–10,000 bp

Before the end of the nineteenth century it was realised that there were two phases of the final withdrawal of glacier ice from Scotland. J. Geikie (1894) believed that the ice sheet was reduced to a series of valley glaciers as the high ground emerged from the ice during deglaciation and that the presence of well-developed morainic mounds (Young 1864) in the valley heads, resting on top of the basal till, indicated that there had been a readvance of ice after the general retreat of the ice sheet. J. Geikie also suggested that the interval between the dissolution of the ice sheet and the readvance of the valley glaciers was one of milder climate. It was some seventy years before the absolute chronology of these two most important phases at the end of the Devensian age began to emerge (Godwin & Willis 1959b; Kirk & Godwin 1963). During the last 20 years the Scottish Lateglacial period has become the most intensively studied 4,000 years of the Quaternary sub-era in Britain (see bibliography in Gray & Lowe 1977).

The earliest interpretations of the environmental changes that took place after the withdrawal of the Devensian ice sheet indicated a three fold sequence linked to the Jessen-Godwin pollen-zonation scheme (Zones I, II and III). There has been considerable debate about the interpretation of pollen Zone I and about the significance and chronology of the boundary between Zones I and II, the latter being equated by some authors with the Alleröd Interstadial of north-west Europe. In some studies in northern Britain (Oldfield 1960; Bartley 1962; Vasari and Vasari 1968; Casseldine 1980), Zone I has been further subdivided into Zone Ia, Ib and Ic with these subzones correlated with the climatic oscillations identified as the equivalent of the Oldest Dryas-Bölling-Older Dryas sequence of north-west Europe. However, Lowe & Gray (1980), in their detailed review of the stratigraphic subdivison of the Lateglacial in northwest Europe, point out (p. 158), ". . .evidence for a distinction between separate Bölling and Alleröd phases is absent from most sites in Europe and is weak, or unsatisfactorily demonstrated, at many sites where the distinction has been made, including Böllingso itself."

One of the main difficulties in palaeoenvironmental reconstructions of the Lateglacial episode has been the conflicting interpretations derived from pollen analysis on the one hand and coleopteran studies on the other (Coope 1975, 1977). Environmental changes during the Lateglacial episode deduced from coleopteran remains do not fit the climatic inferences based on pollen analysis which led to the establishment of the traditional four-fold climatic sequence named Bölling, Older Dryas, Alleröd and Younger Dryas. This conflict is probably explained by the greater ability of the beetle populations to change more rapidly in response to climatic changes while vegetational changes resulting from the same climatic events occur more slowly.

The recognition of the major environmental changes during the Lateglacial episode is based on the interpretation of sediments and their included fossils. Heavy reliance has been placed on pollen grains, macro vegetation-remains, coleoptera and both macro and micro marine fauna (Lowe, Gray & Robinson 1980). Moore (1980, p. 155) has pointed out that, "Although it is tempting to interpret pollen data in climatic terms, the present state of knowledge, both in pollen morphology and in the biogeography and eco-physiology of modern plant populations, is currently very limited. In the light of these gaps in our knowledge a precise analysis of the changing climate of Lateglacial times on the basis of pollen data is still beyond our grasp." Similarly with regard to coleoptera evidence, Lowe & Gray (1980, p. 159), point out that, "Interpretations rest heavily on the assumptons that present (and past) beetle distributions are limited by thermal thresholds. While there are strong argùments for accepting this premise, beetles have not been studied as thoroughly as plant species in recent years, and some modern distributional studies or experimental work are desirable." The use of foraminifers and ostracods in interpreting Lateglacial environments must also be undertaken with great caution. Lord (1980, p. 114), states that "The foraminiferal evidence relating to the Lateglacial deposits cannot be synthesised into a meaningful pattern, as the sites are scattered and most of the dated ones reflect local as much as regional effects. Little is known about how long climatic change . . . takes to affect marine surfaces, or more particularly, bottom waters and how rapidly benthonic organisms react to, for example, sudden fluctuations in temperature and salinity caused by meltwater input." Ostracods have been discussed for over a century in relation to the interpretation of glacial and interglacial sequences (Forbes 1846; Brady, Crossky & Robertson 1874). However, Robinson (1980, p. 121) suggests that there is much work still to be done before ostracods can be reliably used for correlation and palaeoenvironmental reconstruction in the Lateglacial period.

Having clearly stated the limitations of the principal types of evidence used to interpret the dramatic environmental changes that took place in the Lateglacial period in Scotland, the interpretations which follow must be regarded as the best that are possible in the light of present knowledge. It must also be realised that there still remain major problems in dating the environmental changes.

It is for the Lateglacial period that the majority of radiocarbon dates relating to Devensian materials have been obtained. Over 100 such dates are now available and it is tempting to suggest that the chronology of the wastage of the Devensian ice sheet, of the Lateglacial Interstadial and of the Loch Lomond Stadial has been established. However, the probable inaccuracies inherent in the Scottish radiocarbon dates has been clearly demonstrated by Sutherland (1980) and by Lowe & Gray (1980), but Lowe & Gray (1980, p. 165) also point out ". . .it is likely that for some time to come an absolute dating framework for the Lateglacial will rest almost exclusively on radiocarbon dates." The following summary of the chronology of the main environmental changes is therefore only tentative but until the methods of sample collection and analysis are refined and means of eliminating contaminants are found, it is the only local chronology available. The Lateglacial period in Scotland (Table 4.1) can be subdivided into two phases (climatostratigraphic units: Lowe & Gray 1980: Table 2, p. 163):

TABLE 4.1
(Based on Lowe and Gray 1980, p. 163)

Radiocarbon Years bp	Climatostratigraphic Units
	Flandrian Interglacial
10,000 11,000	Loch Lomond (Younger Dryas) Stadial
14,000	Lateglacial Interstadial

Lowe and Gray include transitional periods at each of the three main boundaries but since these are ill-defined and add little to solving the problems associated with the definition of time-transgressive events the more simple scheme is adopted here. There is now strong evidence to suggest that both in Scotland and in north-west Europe as a whole, that the Lateglacial period can be subdivided into two main phases—an interstadial followed by a stadial (Gray & Lowe 1977; Berglund 1979; Lowe & Gray 1980). The interpretation of the environments associated with each of these phases and the dating of the environmental changes which are identified remain debatable.

THE CHRONOLOGY OF THE SCOTTISH LATEGLACIAL

The transition from glacial to non-glacial conditions in Scotland at the end of the Late Devensian ice sheet glaciation can only be dated in very general terms. Figures 4.1, 4.2 and Table 4.2 contain the radiocarbon dates currently available. These dates have been obtained from organic material and shells contained in sediments deposited either in lakes or in the sea after the ice had retreated and therefore provide minimal dates for deglaciation. The oldest date (1a) so far obtained in Scotland for this period is on a shell from the Clyde Beds at Paisley (Fig. 4.2) west of Glasgow (15,625 ± 240). Peacock[33] suggests that this date may have been exaggerated by old carbon and in an additional note states that further shell specimens from the same locality yielded dates of 12,125 ± 210 (outer) and 12,615 ± 230 (inner) but he does not indicate whether these latter shells were taken from the same horizon as the shell producing the very early date.

Rather than debate the validity of any one date it is probably preferable to examine the distribution of all the dates on Figure 4.1. Each date is plotted showing the time range of one standard deviation (i.e. 68 per cent probability that each date falls within the range). Of the 21 dates, seventeen fall within the period 12,200 and 13,200 bp. If these dates are considered in the context of two standard deviations (giving a 95 per cent probability that the 'true dates' fall within these limits) then the group of 17 dates would fall between 11,800 and 13,400 bp. It is obvious that the first three dates on Figure 4.1 (1a, 1b, 1c) all obtained from shells, do not follow the trend demonstrated by all the other dates. The question of the earliest penetration of the sea into the Paisley area still remains unresolved. Perhaps the best indications of the minimum date for deglaciation in Scotland can be obtained from the series of dates (2a, 3a, 1d, 4, 5, 6 and 3b) all of which fall within the range (one standard deviation) of 12,500 and 13,500 bp. There can be little doubt that the Lateglacial Interstadial had begun throughout Scotland by 12,500 bp, but on the limited evidence currently available it may well have begun in some areas before 14,500 bp.

It may be concluded, therefore, that in terms of the evidence currently available from Scotland it is probable that deglaciation occurred sometime before 12,500 bp over most of the country. However, there is evidence from northern England (Pennington 1977a; Coope & Pennington 1977), Sweden (Berglund 1979), and the North Atlantic Ocean (Ruddiman & McIntyre 1973), that the marked thermal improvement associated with the wastage of the Devensian ice sheet occurred between 14,300 and 13,000 bp. Coope (1977, p. 313) states that in Britain

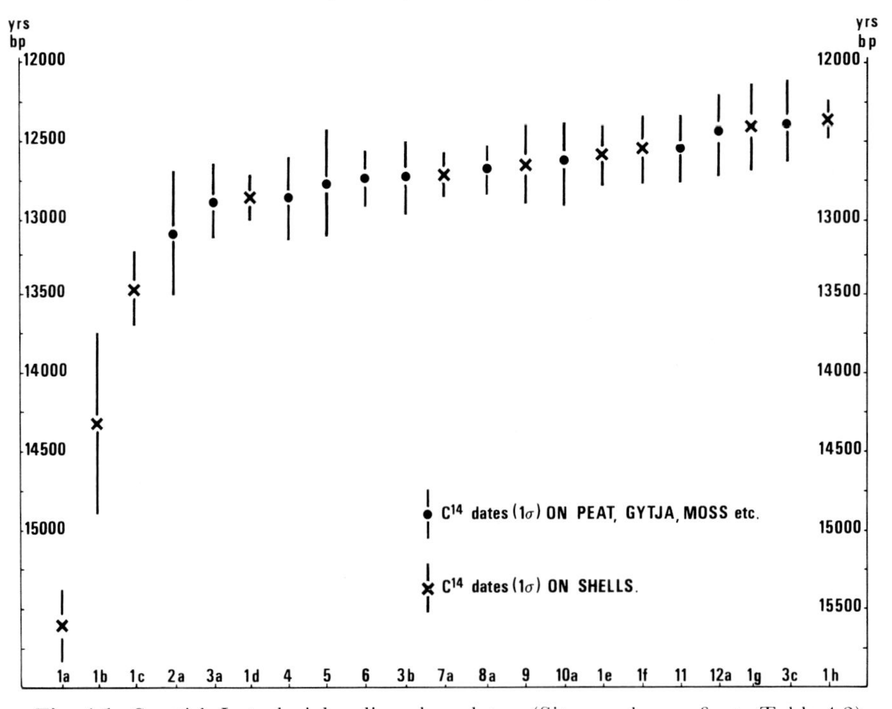

Fig. 4.1. Scottish Lateglacial radiocarbon dates. (Site numbers refer to Table 4.2).

there was a sharp temperature rise "... before 13,000 years ago but later than 14,000 years ago."

It is now widely accepted that after a period of 2,000–3,000 years of interstadial conditions there was a deterioration in Scotland's climate allowing the build up of glacier ice in certain areas of high ground. The period of return to glacial conditions is generally referred to as the Loch Lomond Stadial (Younger Dryas Stadial, generally equated with Zone III of the Jessen-Godwin pollen Zones). The dating of the Loch Lomond Stadial is currently the subject of a great deal of debate (Sutherland 1980; Lowe & Gray 1980; Lowe & Walker 1980a, b). The relevant radiocarbon dates used to define the onset and termination of the Loch Lomond Stadial (Table 4.2, Fig. 4.3) certainly do not indicate a time period that conforms to the Younger Dryas chronozone (11,000–10,000 bp) as defined by Mangerud (1974) and Berglund (1979).

The termination of the Lateglacial Interstadial and the beginning of the Loch Lomond Stadial are extremely difficult to date (Figs. 4.3, 4.4). For example, a sample from Redkirk Point (14b) dated at 11,205 ± 177 bp contained a coleoptera fauna indicative of a decline in the mean July temperature to 12°C but should this

TABLE 4.2
Radiocarbon Dates: Scotland

Date[1]	Lab	Material	Location[2]	Site	Reference
Yrs. bp	no.				
15,625 ± 240	ST3269	Shell	Paisley	1a	Peacock 1971
12,930 ± 160					
$14{,}346\ ^{+625}_{-580}$	SRR923	Shell	Paisley	1b	Brown *et al* 1977.
12,289 ± 310					
13,501 ± 265	SRR925	Shell	Paisley	1c	Brown *et al* 1977.
13,492 ± 330					
13,151 ± 390	SRR304	Gyttja	L. Etteridge	2a	Sissons & Walker 1974.
12,956 ± 240	SRR253	Organic Silt	Cam Loch	3a	Pennington 1975b.
12,922 ± 125	SRR927	Shell	Paisley	1d	Brown *et al* 1977.
12,694 ± 100					
12,940 ± 250	Q643	Wood	Lockerbie	4	Bishop 1963
12,820 ± 350	Q1175	Detrital Mud	L. of Winless	5	Peglar 1979
12,810 ± 155	Q457	Plant Remains	L. Droma	6	Kirk & Godwin 1963.
12,787 ± 185	SRR251	Organic Silt	Cam Loch	3b	Pennington 1977a.
12,765 ± 85	SRR 368	Shell	Lochgilphead	7a	Peacock *et al* 1977.
13,187 ± 85					
12,750 ± 120	HV4989	Gyttja	Callander	8a	Lowe & Walker 1977.
12,720 ± 240	SRR 833	Shell	Geilston	9	Rose 1980.

[1] When two dates are quoted the first is from the 'inner' shell and the second from the 'outer' shell.

[2] A general location name is given and this may include several individual sites which may be identified by reference to the site number/letter and the literature reference.

Date[1]	Lab	Material	Location[2]	Site	Reference
12,710 ± 270	HEL424	Gyttja	Abernethy Forest	10a	Vasari 1977.
12,650 ± 200	Bir 122	Shell	Paisley	1e	Peacock 1971.
13,020 ± 220					
12,610 ± 210	ST3334	Shell	Paisley	1f	Peacock 1971.
12,890 ± 360					
12,596 ± 210	Q1291	Organic Mud	Braemar (Morrone)	11	Switsur & West 1975.
12,510 ± 310	Hel160	Gyttja	Drymen	12a	Vasari 1977
12,464 ± 275	SRR924	Shell	Paisley	1g	Brown *et al* 1977.
13,083 ± 150					
12,436 ± 220	SRR 250	Organic Silt	Cam Loch	3c	Pennington 1977.
12,408 ± 85	SRR-62	Shell	Paisley	1h	Aspen & Jardine 1968.
12,395 ± 195	HV4988	Gyttja	Callander	8b	Walker & Lowe 1977.
12,373 ± 75	SRR482	Shell	Ardyne	13a	Peacock *et al* 1978.
12,649 ± 70					
12,360 ± 85	SRR63	Shell	Lochgilphead	7b	Peacock *et al* 1977.
12,333 ± 110	SRR926	Shell	Paisley	1i	Brown *et al* 1977a
12,389 ± 110					
12,290 ± 250	Q816	Peat	Redkirk Pt.	14a	Bishop & Coope 1977.
12,251 ± 250	Q1078	Detrital Mud	Din Moss	15	Switsur & West 1973.
12,210 ± 100	SRR 369	Shell	Lochgilphead	7c	Peacock *et al* 1977.
12,729 ± 850					
12,092 ± 190	SRR 367	Shell	Lochgilphead	7d	Peacock *et al* 1977.
12,066 ± 70	SRR489	Shell	Lochgilphead	7e	Peacock *et al* 1977.
12,151 ± 60					
12,060 ± 320	HEL161	Gyttja	Drymen	12b	Vasari 1977.
11,900 ± 260	Hel417	Gyttja	Loch of Park	16a	Vasari 1977.
11,805 ± 205	ST3267	Shell	Dumbarton	17	Peacock 1971.
11,805 ± 180	ST3263	Shell	L. Creran	18a	Peacock 1971.
11,800 ± 170	I2234	Shell	Lake of Menteith	19	Sissons 1967a
11,770 ± 87	SSR305	Gyttja	L. Builg	20	Clapperton *et al* 1975.
11,700 ± 170	I2235	Shell	Drymen	12c	Sissons 1967a.
11,680 ± 165	SRR-481	Shell	Ardyne	13e	Peacock *et al* 1978.
12,150 ± 115					
11,653 ± 190	SRR 696	Peat	Glasgow	21a	Dickinson *et al* 1976.
11,530 ± 210	ST3332	Shell	L. Creran	18b	Peacock 1971.
11,300 ± 300					
11,527 ± 551	SRR-483	Shell	Ardyne	13c	Peacock *et al* 1978.
11,470 ± 70					
11,520 ± 220	HEL418	Gyttja	L. Kinord	22a	Vasari 1977.
11,520 ± 250	HAR-931	Shell	Helensburgh	23	Rose 1980.
11,480 ± 440	HEL174	Gyttja	L. Kinord	22b	Vasari 1977.
11,430 ± 220	ST3306	Shell	L. Creran	18c	Peacock 1971.
6,705 ± 130					
11,385 ± 290	HV4987	Gyttja	Callander	8c	Lowe & Walker 1977.
11,350 ± 285	GU1084	Gyttja	Rannoch Moor	24a	Lowe & Walker 1980.
11,330 ± 170	I5308	Shell	L. Spelve	25	Gray & Brook 1972.

Date[1]	Lab	Material	Location[2]	Site	Reference
11,313 ± 140 11,694 ± 130	SRR365	Shell	Lochgilphead	7f	Peacock *et al* 1977.
11,300 ± 245	GU1083	Gyttja	Rannoch Moor	24b	Lowe & Walker 1980.
11,290 ± 165	SRR303	Gyttja	L. Etteridge	2b	Sissons & Walker 1974.
11,260 ± 240	HEL-423	Gyttja	Abernethy Forest	10b	Vasari 1977.
11,210 ± 190	SRR756	Moss	North Kerse	26	Dickson *et al* 1976.
11,210 ± 150	Birm668	Peat	Glasgow	21b	Dickson *et al* 1976.
11,205 ± 177	Birm41	Peat	Redkirk Point	14b	Bishop & Coope 1977.
11,196 ± 130 10,962 ± 115	SRR366	Shell	Lochgilphead	7g	Peacock *et al* 1977.
11,140 ± 110	SRR761	Peat	Glasgow	21c	Dickson *et al* 1976.
11,116 ± 110 11,165 ± 70	SRR485	Shell	Ardyne	13d	Peacock *et al* 1978.
11,115 ± 220	Q1267	Organic Silt	Abernethy Forest	10c	Birks & Matthewes 1978.
10,975 ± 110 11,279 ± 125	SRR364	Shell	Lochgilphead	7h	Peacock *et al* 1977.
10,898 ± 127	Birm 40	Peat	Redkirk Point	14c	Bishop & Coope 1977.
10,838 ± 100 10,232 ± 210	SRR484	Shell	Ardyne	13e	Peacock *et al* 1978.
10,800 ± 230	Q104	Peat	Garral Hill	27	Godwin & Willis 1959.
10,770 ± 90	HV5644	Gyttja	Amulree	28a	Lowe & Walker 1977.
10,764 ± 120	SRR-302	Gyttja	L. Etteridge	2c	Sissons & Walker 1974.
10,752 ± 135	SRR615	Shell	Ardyne	13f	Peacock *et al 1978.*
10,698 ± 490	SRR-249	Organic Silt	Cam Loch	3d	Pennington 1977a.
10,670 ± 85	HV5647	Gyttja	Callander	8d	Lowe & Walker 1977.
10,660 ± 240	SRR1074	Gyttja	Rannoch Moor	24c	Lowe & Walker 1980.
10,640 ± 260	HEL419	Gyttja	L. Kinord	22c	Vasari 1977.
10,617 ± 63 10,553 ± 54	SRR486	Shell	Ardyne	13g	Peacock *et al* 1978.
10585 ± 450	SRR248	Organic Silt	Cam Loch	3e	Pennington 1977a.
10,520 ± 330	BIRM 723	Gyttja	Rannoch Moor	24d	Lowe & Walker 1980.
10,480 ± 150	HV5646	Gyttja	Callander	8e	Lowe & Walker 1977.
10,420 ± 230	BIRM 722	Moss	Rannoch Moor	24e	Lowe & Walker 1980.
10,420 ± 160	HV4985	Gyttja	Callander	8f	Lowe & Walker 1977.
10,400 ± 100	K1865	Gyttja	Murraster	29	Johansen 1975.
10,390 ± 200	BIRM 858	Organic Mud	Rannoch Moor	24f	Lowe & Walker 1980.
10,370 ± 290	BIRM 724	Organic Mud	Rannoch Moor	24g	Lowe & Walker 1980.
10,300 ± 185	Q815	Wood	Redkirk Pt.	14d	Bishop & Coope 1977.
10,280 ± 220	HEL416	Gyttja	Loch of Park	16b	Vasari 1977.
10,270 ± 100	HV5646	Gyttja	Amulree	28b	Walker & Lowe 1977.
10,254 ± 220	Q955	Diatomaceous Mud	L. Coir	30	Vasari 1977.
10,230 ± 220	HEL422	Gyttja	Abernethy Forest	10d	Vasari 1977.
10,226 ± 190	SRR-247	Organic Mud	Cam Loch	3f	Pennington 1975b.
10,160 ± 200	Birm 859	Organic Mud	Rannoch Moor	24h	Lowe & Walker 1980.
10,100 ± 135	HV5650	Organic Silt	Blackness	31	Walker 1975.
10,060 ± 270	HEL504	Gyttja	L. Cuithir	32	Vasari 1977.

Date[1]	Lab	Material	Location[2]	Site	Reference
10,010 ± 220	HEL420	Gyttja	L. Kinord	22d	Vasari 1977.
10,010 ± 230	HEL162	Gyttja	Drymen	12d	Vasari 1977.
9,820 ± 250	HEL421	Gyttja	L. Kinord	22e	Vasari 1977.
9,800 ± 160	Birm 854	Organic Mud	Rannoch Moor	24i	Lowe & Walker 1980.
9,740 ± 170	Q1268	Organic Mud	Abernethy Forest	10e	Birks & Matthewes 1978.
9,730 ± 180	GU1101	Organic Mud	Rannoch Moor	24j	Lowe & Walker 1980.
9,697 ± 90	SRR1072	Peat	Rannoch Moor	24k	Lowe & Walker 1980.
9,595 ± 215	GU1100	Organic Mud	Rannoch Moor	24l	Lowe & Walker 1980.
9,490 ± 160	HV5649	Gyttja	Blackness	33	Walker 1975a.
9,440 ± 310	Birm 855	Gyttja	Rannoch Moor	24m	Lowe & Walker 1980.
9,440 ± 70	SRR1419	Gyttja	Rannoch Moor	24n	Lowe & Walker 1980.
9,405 ± 260	SRR 301	Gyttja	L. Etteridge	2d	Sissons & Walker 1974.
9,365 ± 120	HV5645	Gyttja	Callander	8g	Lowe & Walker 1977.
9,260 ± 100	HV4984	Gyttja	Callander	8h	Lowe & Walker 1977.
9,152 ± 95	SR1073	Peat	Rannoch Moor	24,o	Lowe & Walker 1980.
9,115 ± 120	HV5642	Gyttja	Amulree	28c	Lowe & Walker 1977.
8,920 ± 80	SRR1418	Gyttja	Rannoch Moor	24p	Lowe & Walker 1980.
8,340 ± 160	Birm 856	Gyttja	Rannoch Moor	24q	Lowe & Walker 1980.
8,130 ± 40	SRR1417	Gyttja	Rannoch Moor	24r	Lowe & Walker 1980.
8,120 ± 140	Birm 857	Organic Mud	Rannoch Moor	24s	Lowe & Walker 1980.
8,040 ± 50	SRR1416	Organic Mud	Rannoch Moor	24t	Lowe & Walker 1980.

[1] When two dates are quoted the first is from the 'inner' shell and the second from the 'outer' shell.

[2] A general location name is given and this may include several individual sites which may be identified by reference to the site number/letter and the literature reference.

be considered an Interstadial deposit or an early phase of the Loch Lomond Stadial? If one accepts individual authors' published classifications of samples then, the Loch Lomond Stadial began sometime between 11,300 (10c) and 10,600 (27) on the basis of one standard deviation of the dates from these sites. Similarly the youngest Loch Lomond Stadial deposits so far dated fall in the range of 10,250 to 9,750 bp (32, 22d, 12d).

The most detailed work relating to the definition of a Lateglacial chronostratigraphic boundary within the Scottish Lateglacial has been done with reference to the end of the Loch Lomond Stadial and the beginning of the Flandrian in the Rannoch Moor area (Lowe & Gray 1980; Lowe & Walker 1980a, b). Sutherland (1980) has discussed in detail the possible causes of contamination of the material used in determining the end of the Stadial. These include contamination by older and younger carbon by isotopic fractionation and the thickness of the dated sediment slices required to produce a radiocarbon date. Some of the samples used to produce dates from the Scottish Lateglacial were 30–100 mm thick and such dates will relate to the average age of the whole sample and will therefore be younger than the initial time of organic deposition. Rates of sedimentation have been estimated for five sites by Sutherland and they range from 0·17 to 0·67 mm/yr. These figures suggest that dates may be 50 to 200 years younger than the initial time of sediment accumulation. There is also growing concern that hard-water errors may be a significant problem in dating samples from freshly deglaciated areas and that the late melt out of glacier ice from some of the kettle holes may produce a further error of at least two hundred years. Bearing in mind all these possible errors it is not surprising that the earliest "Flandrain" dates (i.e. materials indicative of the termination of the Loch Lomond Stadial at any site) vary between 11,600 and 8,000 bp. Even when considering only the dates from the Rannoch Moor area, the earliest Flandrian dates at any one site (one standard deviation) range from 11,600 (24a) to 8,090 (24r).

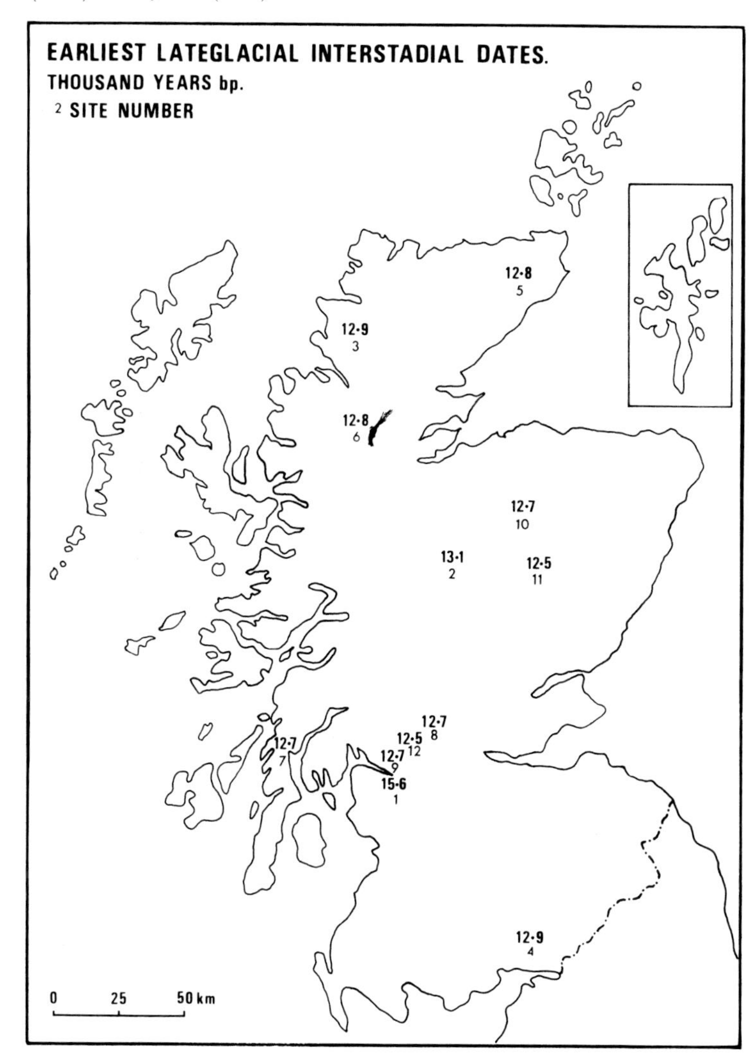

Fig. 4.2.

Earliest Lateglacial Interstadial radiocarbon dates.

Even if one accepts the published dates (Fig. 4.1) at their face value the most pessimistic and probably the most realistic conclusions which can be reached about the chronology of the environmental changes which took place in Scotland following the retreat of the Late Devensian ice sheet are as follows. The onset of the Lateglacial Interstadial occurred some time before 12,500 bp and that the Lategla-

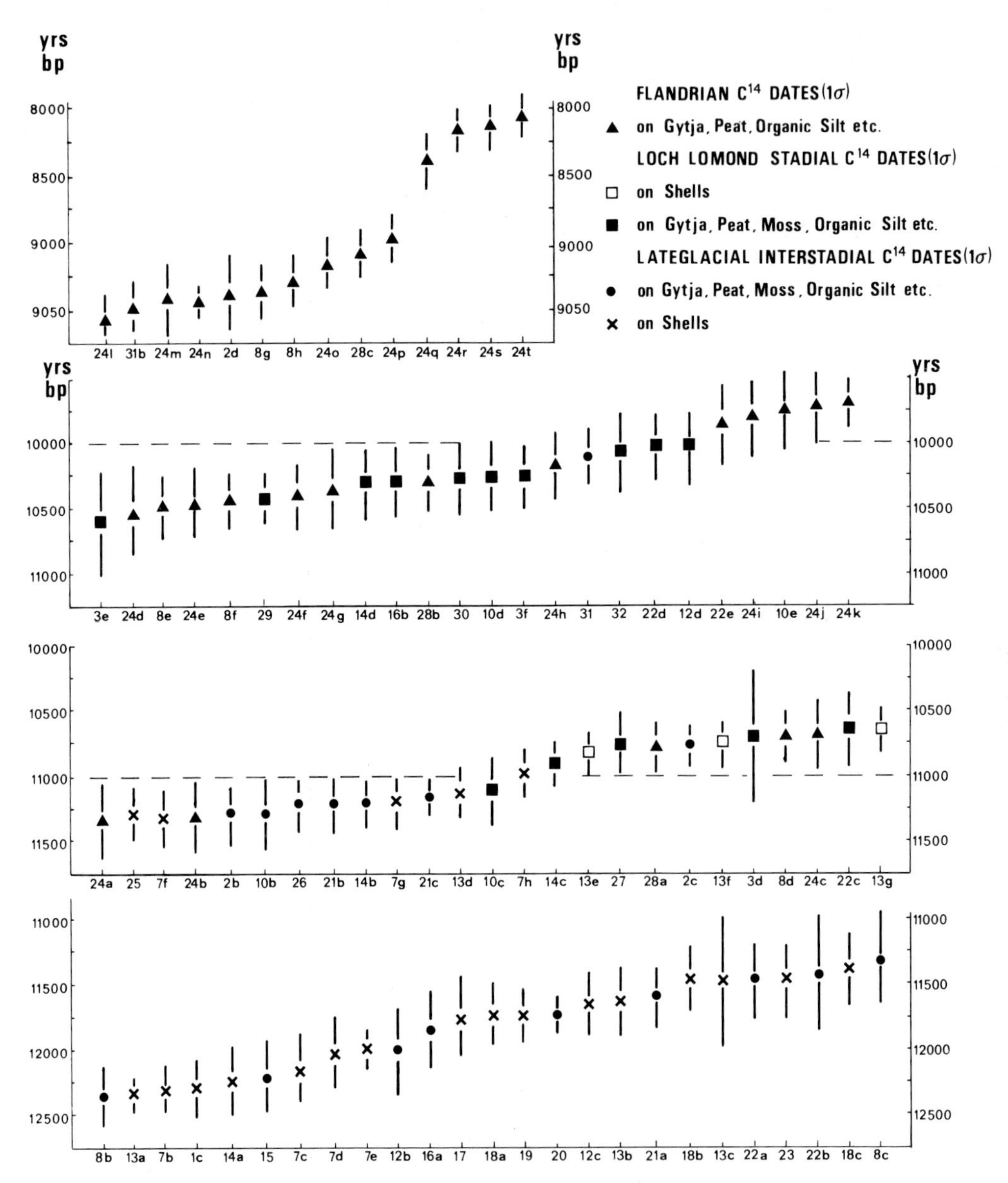

Fig. 4.3. Scottish Lateglacial and early Postglacial readiocarbon dates. (Site numbers refer to Table 4.2).

cial Stadial revertence took place sometime between 11,600 and 8,000 bp. On the basis of the published dates some of the oldest Flandrian dates are older than some of the oldest Loch Lomond Stadial dates. This is, of course, not unreasonable in that the environmental changes being dated would take place at different times at different places. It is disconcerting however when such overlaps occur within a few kilometres, as is the case with some of the Rannoch Moor sites.

It must be admitted, therefore, that radiocarbon dating as presently practiced is unsuitable for dating faunal, floral and by implication, climatic changes in the Scottish Lateglacial. The technique is simply not sufficiently accurate to allow climatostratigraphic boundaries to be dated. If the major environmental changes took place over a period of about 1,000 years, then a technique which may include

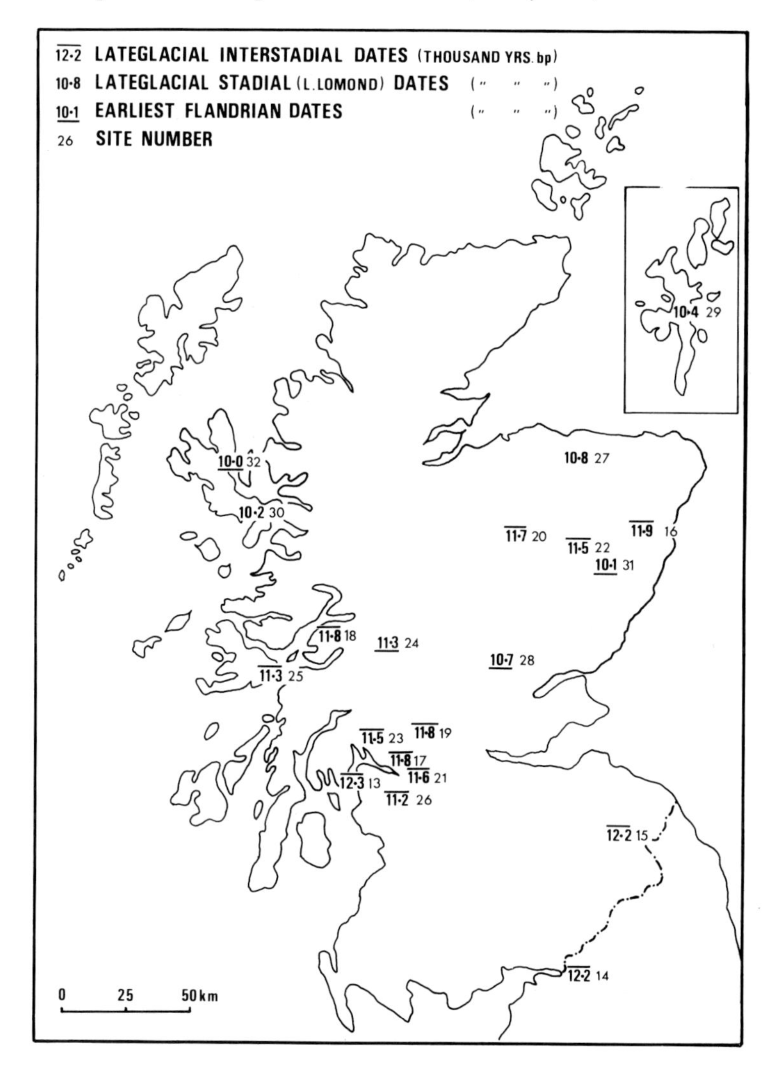

Fig. 4.4.

Radiocarbon dates indicative of the onset and termination of the Loch Lomond Stadial.

errors of 500 years as a result of contamination or sampling procedures and statistical errors of ±200 years at one standard deviation, is unsuitable for providing a chronological framework for these changes. If radiocarbon dates are only accepted at the 95 per cent significance level (two standard deviations) then any published date of say 11,000 bp (Fig. 4.5) may be in error by 500 years as a result of contamination and sampling procedures and should really be expressed as a range between 11,900 and 10,100 bp (i.e. 11,500 and 10,500 plus or minus two standard deviations of 200 years). It seems necessary to conclude, therefore, that the absolute chronology and therefore the rate of change of environmental conditions in the Lateglacial period in Scotland have not yet been established and such deductions will have to await either the refinement of the radiocarbon technique or the development of other absolute dating techniques.

It may be concluded, therefore, that purely on the basis of radiocarbon dates of samples from Scottish sites that sometime between 11,900 and 10,800 bp (23 out of 46 dates fall within the period if one standard deviation is plotted) there was a revertence to glacial/periglacial conditions in Scotland. This stadial came to an end in Scotland sometime between 10,900 and 9,300 bp (16 out of 29 dates fall within this period if one standard deviation is plotted).

In the light of the above discussion it would seem wise to make any classification of the Lateglacial period as general as possible and, until such time as dating techniques are improved and the nature of the time-transgressive changes in the environment are better understood, to regard the chronology of these changes only in very general terms. There certainly does not appear to be one single site in Scotland at which it is possible to establish reliable chronozone boundaries which can be used for correlation with other sites in Scotland or other parts of Britain or Europe.

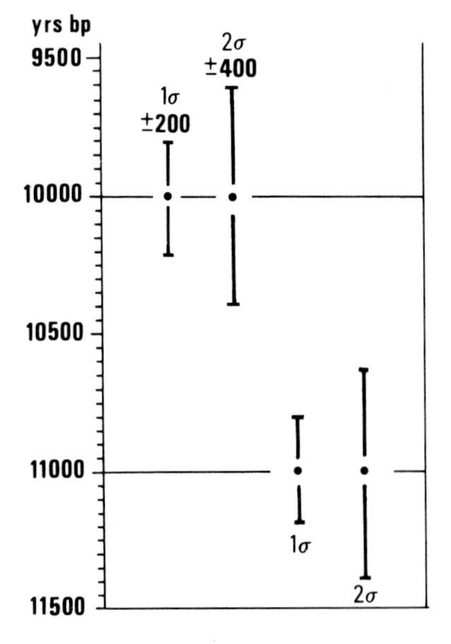

Fig. 4.5.

The resolution of typical radiocarbon dates at one and two standard deviations of 200 years.

THE LATEGLACIAL INTERSTADIAL: circa 14,000–11,000 bp

There are some 40 sites in Scotland (Table 4.2, Fig. 4.6) at which sediments relating to the Lateglacial interstadial have been investigated. These sediments are mainly of two types: lacustrine and marine. The most common sediments are those which accumulated in kettle holes or rock basins and therefore consist of clays, silts and sands in which pollen grains, diatoms, macro vegetation fragments and coleoptera have been preserved. At the same time as these lacustrine sediments were accumulating, sea level was relatively high around the Scottish shoreline and there are extensive marine deposits which were laid down then and which subsequently have been uplifted above present sea level. These marine sediments contain both micro and macro fauna.

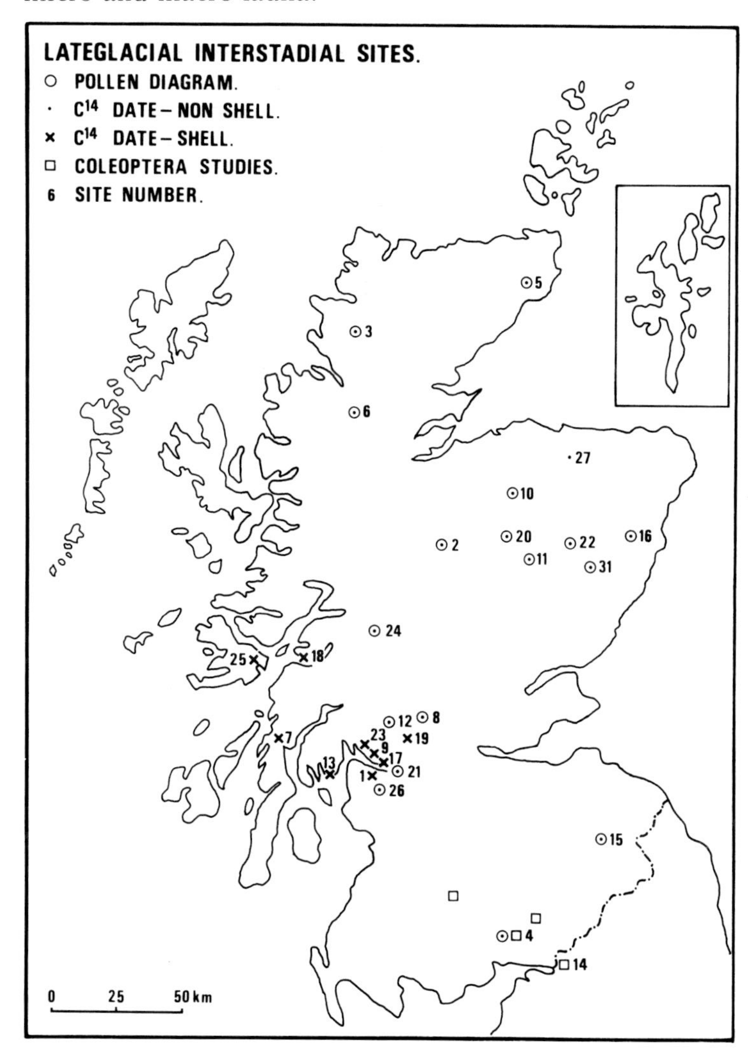

Fig. 4.6.

Lateglacial Interstadial sites. (Site numbers refer to Table 4.2).

It is from the biogenic material in these lacustrine and marine sediments that information about the vegetation, soils and fauna of the Lateglacial Stadial has been obtained. The same sediments also provide some information about the type and rate of geomorphic processes operating at that time. By analogy with present day species distribution of both the fauna and vegetation represented by the Lateglacial fossils it is possible to suggest the character of the climate prevailing in Scotland at the time of the Lateglacial Interstadial.

The distribution of investigated Lateglacial Interstadial sites in Scotland (Fig. 4.6) is surprisingly uniform. The only major areas lacking in any information are the central and eastern Southern Uplands, the Inner Hebrides (except Skye) and the Outer Hebrides. It is not surprising that there are no sites in the Western Highlands as this area was covered by ice during the Loch Lomond Stadial (Fig. 4.12) and the deposits of the interstadial were either removed by glacial erosion or buried by subsequent glacial deposits.

The majority of the sites on Figure 4.6 have provided information for the construction of pollen diagrams. Only four sites have provided fossil coleoptera. It is tempting to suggest that at least the large number of pollen sites would allow the reconstruction of the vegetation of the Lateglacial Interstadial. However, it must be remembered that each site has its own local environment and that there are major regional differences in Scotland to-day as a result of altitude, exposure and soil conditions. No attempt will therefore be made to discuss detailed regional variations in Scotland during the Lateglacial Interstadial. The country will be discussed as a whole and only where data for an individual site or group of sites indicate marked differences from the general pattern will these local characteristics be discussed.

LANDFORMS AND SEA-LEVELS

The landforms and deposits produced by the Devensian ice sheet in Scotland were discussed in Chapter 3. It has been suggested that ice wastage was probably very rapid and this is supported by evidence obtained outwith Scotland which suggests a rapid amelioration of climate about 13,500 bp (Ruddiman & McIntyre 1973; Coope 1977). It seems likely therefore that the whole of Scotland, or at least very large parts of it, were deglaciated in a period lasting only three or four thousand years. The newly exposed surfaces of glacial and fluvioglacial erosion and deposition were immediately subject to processes of slope modification by frost action, mass movement and fluvial, aeolian and marine processes.

There is little evidence of intense periglacial action during the closing phases of the Devensian glaciation but it is highly likely that such processes did occur on the higher ground when 'dead ice" still remained in the valleys. Evidence of such periglacial modification would most likely be destroyed either by the very active fluvial processes of the Lateglacial Interstadial or by the glacial and periglacial activity associated with the subsequent Loch Lomond Stadial.

The development of the drainage system and associated slope processes during and immediately after deglaciation probably produced rapid and spectacular changes in the glacial and fluvioglacial landforms (Price 1980a, p. 91). With little or no vegetation cover the steeper slopes would be subject to erosion. Valley-floors filled with glacial and fluvioglacial deposits would be incised and terraces produced. Such action would be intensified as isostatic uplift consequent upon deglaciation began to be more important than the eustatic rise of sea level. The Lateglacial Interstadial in most of Scotland was characterised by a falling relative sea-level and therefore fluvial incision was a major modifier of the landscape at that time.

Much has been written about shoreline altitudes subsequent to deglaciation in Scotland (Sissons & Smith 1965; Synge & Stephens 1966; Cullingford & Smith 1966; Smith, Sissons & Cullingford 1969; Sissons 1974a, 1976a; Armstrong *et al* 1975; Peacock 1975, 1977, 1978; Jardine 1977, 1979, 1982; J. S. Smith 1977; Synge 1977; Gray & Lowe 1977; Price 1980b). Although a great deal of information is known about the altitude of shoreline fragments, their distribution and their probable gradients resulting from isostatic uplift, there is little reliable information about the chronology of the invasion by the sea of the newly deglaciated land areas. The maximum marine incursion in the Lateglacial Interstadial occurred in the east coast firths (Fig. 4.7), towards the northern end of the Great Glen and in the sea-lochs of the Firth of Clyde. These high relative sea levels were closely associated with the retreating margin of the Devensian ice sheet as their shorelines can be traced inland to the margins of glacial outwash spreads containing kettle holes. The abrupt termination of some of these high shorelines at inland sites can only be explained by the contemporaneous presence of an ice margin. Sissons (1976, p. 122) describes a shoreline (Main Perth) which, east of Stirling, attains an altitude of 38 m but he goes on to point out that immediately west of Stirling the highest shoreline does not exceed 20 m. He states that "This pronounced drop in the marine limit suggests a halt in ice retreat or possibly a minor readvance." Similar sharp drops in the marine limit have been recorded in Loch Fyne and Loch Long by Sutherland (Gray & Lowe 1977, p. 169). Until such time as either the actual date of deglaciation or the age of the marine deposits forming these high level shorelines are known, little can be said about the position of the coastline of Scotland at a particular time in the Lateglacial Interstadial. However, there is no doubt that after the maximum marine limit was reached, relative sea level began to fall rapidly and by the onset of the Loch Lomond Stadial sea level was not very different from that of to-day near Grangemouth on the Firth of Forth and in the Firth of Clyde.

On the west coast, there is a substantial body of evidence to indicate a high stand of relative sea level during ice front retreat in the Inner Hebrides and the sea lochs of the Firth of Clyde. Penetration of the sea into these areas was dependent on the rate of rise of sea level on a global scale as the ice sheets wasted and the rate of recovery of the land masses as a result of the glacial unloading. Investigations in the Solway Firth (Pantin 1975), the Firth of Clyde (Deegan *et al* 1973), the Malin Sea area (Dobson & Evans 1975) and the Sea of the Hebrides (Binns *et al* 1973) by the Institute of Geological Sciences indicate that there are marine sediments overlying the glacial and fluvioglacial sediments of the Devensian ice sheet. In the Solway Firth the glacial till is succeeded by lagoonal deposits which are believed to have accumulated in a water body connected with the open sea, but colder and less saline than the present waters of the Irish Sea. The overlying marine beds are thought to range from Late Devensian to Holocene in age. The total sequence suggests the proximity of marine waters to the Solway Firth area at a level below present sea-level while the Late Devensian ice front lay in the vicinity or a short distance to the north of the present northern shore of the Solway Firth. Jardine (1971, p. 103) concludes that it is doubtful if Late Devensian shorelines ever existed in the Solway Firth at altitudes above present sea-level. This conclusion is supported by a date from freshwater peat on the foreshore at Redkirk Point of 12,290 ± 250 bp (Bishop & Coope 1977).

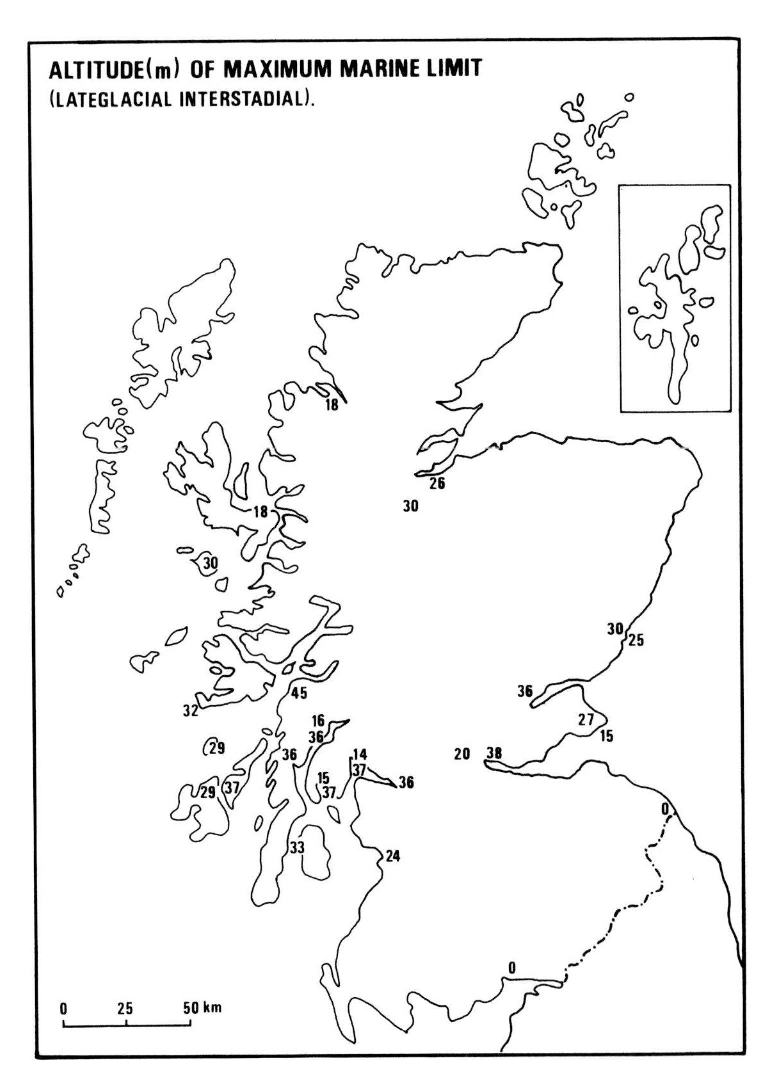

Fig. 4.7.

Altitude (metres) of the Lateglacial maximum marine limit.

In the Firth of Clyde, Late Devensian marine deposits occur at numerous localities at levels between a few metres below O.D. and c. 37 m (Fig. 4.8). Around Paisley these marine deposits are known to attain a thickness of over 20 m. At Wester Fulwood, near Paisley a date of 12,650 ± 200 bp obtained from large valves of *Arctica islandica* from Clyde Beds at 3 m O.D., indicates the age of the penetration of the Late Devensian Interstadial sea into this area.

As early as 1874 Brady, Crosskey and Robertson demonstrated that the oldest fossil ostracod assemblages from the Clyde Beds of the Paisley area were associated with what they described as Arctic-Sea-water temperatures. Three species which Brady, Crosskey and Robertson recognised as indicating these cold water conditions: *Krithe glacialis*, *Cytheropteron montosiensis* and *Rabilimis mirabilis*, are found alive and breeding to-day no further south than the waters of the Barents Sea or the

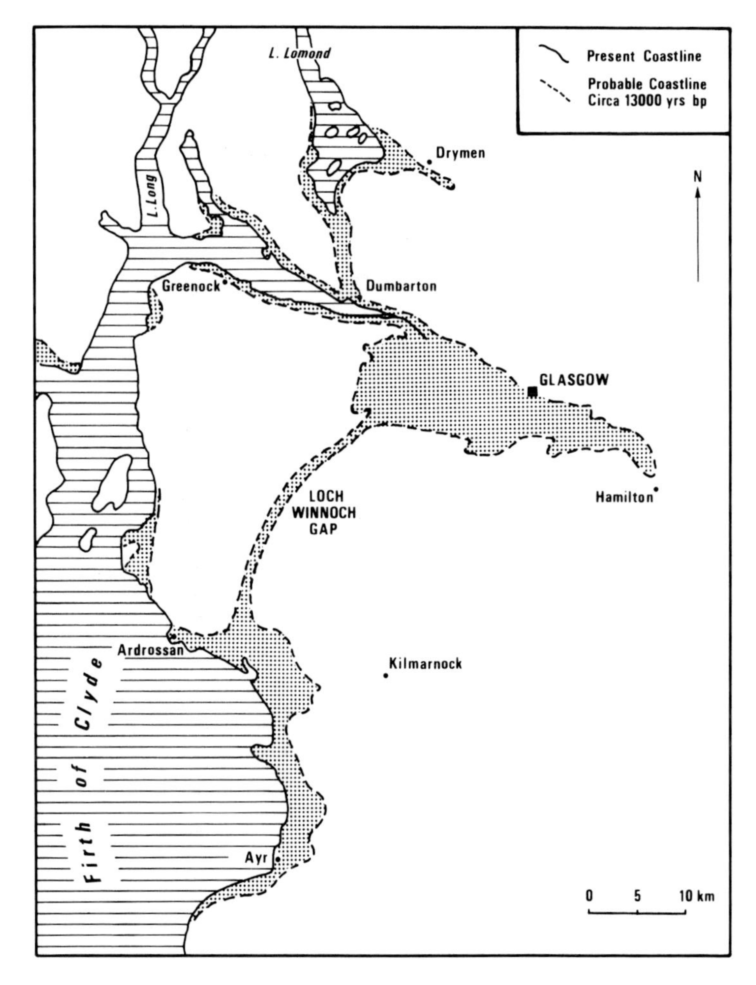

Fig. 4.8.

Lateglacial coastline of the Firth of Clyde.

fjords of East Greenland north of 76°N (Hazel 1967, 1970). These fossils relate to deposits accumulated immediately after deglaciation of the Firth of Clyde, in waters with temperatures close to freezing point and turbid with suspended sediment (Robinson 1980). Comparable deposits and fossils occur at Errol on the Firth of Tay but there appears to be significant differences between the Clyde Beds and the Errol Beds, and Peacock (1981) has suggested that these differences may relate to the fact that the Errol Beds pre-date 13,500 bp, whereas the Clyde Beds accumulated mainly between 13,500 and 10,000 bp. The Errol Beds contain a fauna of low diversity and of mid to high Arctic aspect.

There is uncertainty about the route by which the Late Devensian sea penetrated the Dumbarton-Glasgow area. Entry may have been through the Lochwinnoch Gap (Fig. 4.8), as was suggested originally by Peacock (1971) and supported by Rose (1980) on the basis of evidence present at Geilston, or it may have been via Greenock eastwards along the present course of the Clyde Estuary (Sissons 1974b). By about 11,700 bp, however, the sea had encroached as far as the present coast and in some areas, for example around Paisley, well beyond its present limits. In addition, the presence of exposures of Clyde Beds on the western side of Loch Lomond together with shells radiocarbon dated at 11,700 ± 170 bp in glacial deposits in the Loch Lomond basin laid down by the Lateglacial Stadial glacier, suggests that at least the southern end of the Loch Lomond valley was penetrated by the sea in the Lateglacial Interstadial.

The altitude of the marine limit in the Glasgow area during the Lateglacial Interstadial is a matter of debate. Rose (1975) and Brown (1980) suggest altitudes of 34 m O.D. and 36 m O.D. respectively, but Brown also suggests that sands and gravels of "probable marine origin" (Brown 1980, p. 13) up to 41 m O.D. occur on the north side of Glasgow at Dougalston. In the Firth of Clyde the highest marine limit recorded on the Isle of Arran is 33 m at Imachar Point (Gemmell 1973), on the west coast of the island, and 36 m in Loch Fyne and 40 m in Loch Long (Sutherland, in Gray & Lowe, 1977).

Some of the most spectacular landforms produced by the Lateglacial high sea level are to be found on the west coast of Jura (Pl. 4.1). In West Loch Tarbert massive pebble and boulder beaches (McCann 1964) relate to a sea level which dropped progressively from c. 37 m to c. 17 m as the land uplift outpaced sea level rise. Similar features have been described on Colonsay by Jardine (1977) at an altitude of 29 m and on the island of Rhum at altitudes between 30 and 12 m by McCann and Richards (1969).

The early phases of the relatively high sea-level recorded at various localities on the Firth of Clyde and in the Inner Hebrides presumably relate to the early stage of deglaciation sometime after 16,000 bp and before 12,500 bp when the Clyde Beds were being deposited. Because of the steep nature of much of the coastline affected by this high sea-level, evidence of its former existence is limited. Apart from the major embayment around Paisley and limited incursions between Ardrossan and Ayr and near Girvan the position of the coastline (Fig. 4.8) would not have been radically different from that of the present. The only extensive area known to have

A.

B.

Plate 4.1.

A. Raised beach gravels on the south side of West Loch Tarbet, Jura.
B. Raised marine platforms—Kincraig Point, Fife.

been invaded by the Lateglacial Interstadial sea further north was around Crinan and Lochgilphead (Fig. 4.9). Peacock *et al* (1977) state that, "Prior to approximately 12,500 bp Loch Gilp was connected by a strait to the Sound of Jura." They also state that, ". . .sea-level was about 25–30 m O.D. at 12,500 bp falling to about 10 m O.D. at 12,000 bp and remaining between this level and 4 m O.D. until sometime after 11,000 bp." On the basis of the included fossils in the marine deposits in the Lochgilphead area, Peacock *et al* (1977, p. 26) deduce that, "During the period, which ended about 11,000 bp, sea water temperatures were consistently lower than at present, and for the 12,000–11,000 bp interval, high Boreal to low Arctic conditions are inferred."

Little detailed evidence is available about Lateglacial Interstadial sea-levels in the north of Scotland. Sissons (1967, p. 162 and Fig. 68) states that, "The deposits [of the Lateglacial Interstadial sea] do not occur above present sea-level in the Orkneys and Shetlands and appear to be similarly absent from part of Caithness. On the opposite coast the upper limit falls northwards from Wester Ross through Sutherland and descends towards the Outer Hebrides."

There are not yet sufficient data to construct isobases for the maximum marine transgresson during the Lateglacial Interstadial. The earliest shorelines to be formed on the east coast were those in Fife (Cullingford & Smith 1966) which

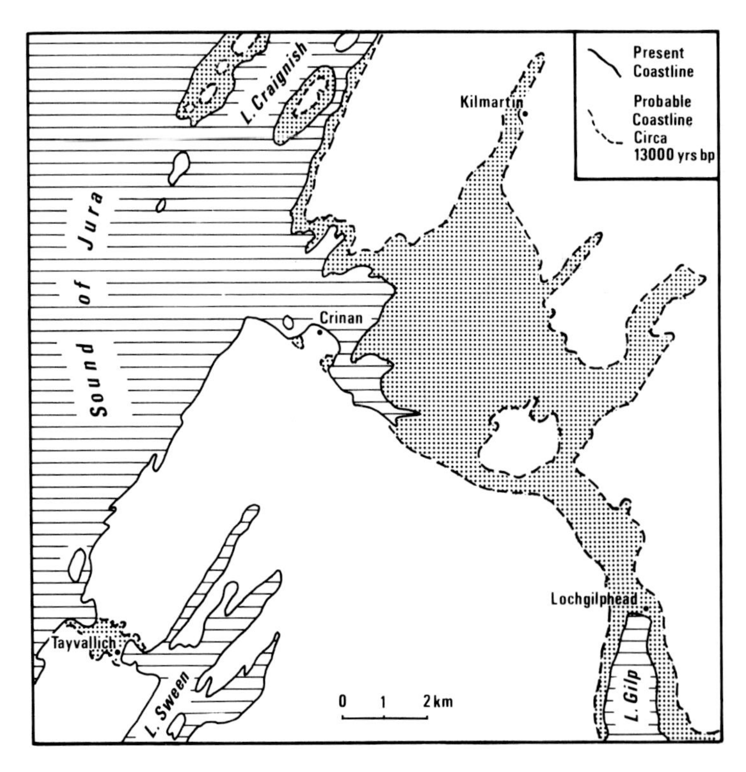

Fig. 4.9.

Lateglacial coastline between Crinan and Loch Gilphead.

have gradients ranging from 1·26 m/km for the oldest to 0·60 m/km for the youngest. As ice wastage continued several other raised shorelines were produced, the best developed of which has been labeled as the Perth shoreline (Sissons & Smith 1965; Cullingford 1977) and that shoreline has a gradient of 0·43 m/km and a direction E17°S. On the west coast it is likely that the earliest shorelines were developed in the lower Firth of Clyde and in the Inner Hebrides on Jura, Islay and Colonsay.

Until such time as the various Lateglacial Interstadial shorelines can be dated there is little hope of successfully correlating those which have been recognised at various locations on both the west and east coasts. The period of rapid deglacierisation which took place sometime around 14,000 bp was characterised by cold seas which produced benches, beaches, deltas and bottom deposits which, because of the isostatic depression of the land at the time of their formation, are now found at altitudes up to 40 m above present sea-level.

VEGETATION AND SOILS

There is little direct evidence from Scotland to suggest how long it took for soils to develop on, and vegetation to colonise, the newly deglaciated terrain. Based on knowledge of areas recently deglaciated in Alaska and Iceland it is probable that within a few decades or at most a few hundred years of the last ice melting in an area the first plants would arrive and the processes of soil formation would be active. The availability of several dozen pollen profiles, some of which have been radiocarbon dated, from various parts of Scotland suggest that the initial vegetation type of the Lateglacial Interstadial was an open habitat consisting primarily of grasses (*Gramineae*), sedges (*Cyperaceae*) and mosses, such plants being the colonisers of the newly developed mineral soils. The open habitat, treeless, pioneer vegetation developed under a temperate climatic regime and as time passed it developed in some areas into a heath, dwarf shrub and grassland association with some juniper, willow and birch. The general tree line (birch) did not reach as far north as Scotland during the Lateglacial Interstadial.

As a result of detailed work in the Eastern Grampians (Fig. 4.10A), Lowe & Walker (1977, p. 116) concluded that the landscape "...ranged from floristically-rich closed or semi-closed grassland, with juniper, willow heath, dwarf birch and occasional copses of tree birch in sheltered localities, to grass and moss heath communities interspersed with patches of bare and unstable soils on the upper slopes and plateau surfaces."

In lowland Aberdeenshire (Vasari & Vasari 1968; Vasari 1977) an open vegetation dominated by *Rumex* was replaced by a more closed vegetation of *Empetrum* heath which in turn was succeeded by a *Juniperus* heath with some scattered birch (*Betula*) trees (Fig. 4.10B). In the north-west Highlands Pennington (1977) describes a similar heath-grassland-dwarf shrub association and concludes that at no time in the Lateglacial Interstadial did birch woodland develop.

Evidence from other Scottish Lateglacial pollen diagrams (Donner 1957; Kirk & Godwin 1963; Newey 1970; Moar 1969a; Pennington & Lishman 1971) agrees

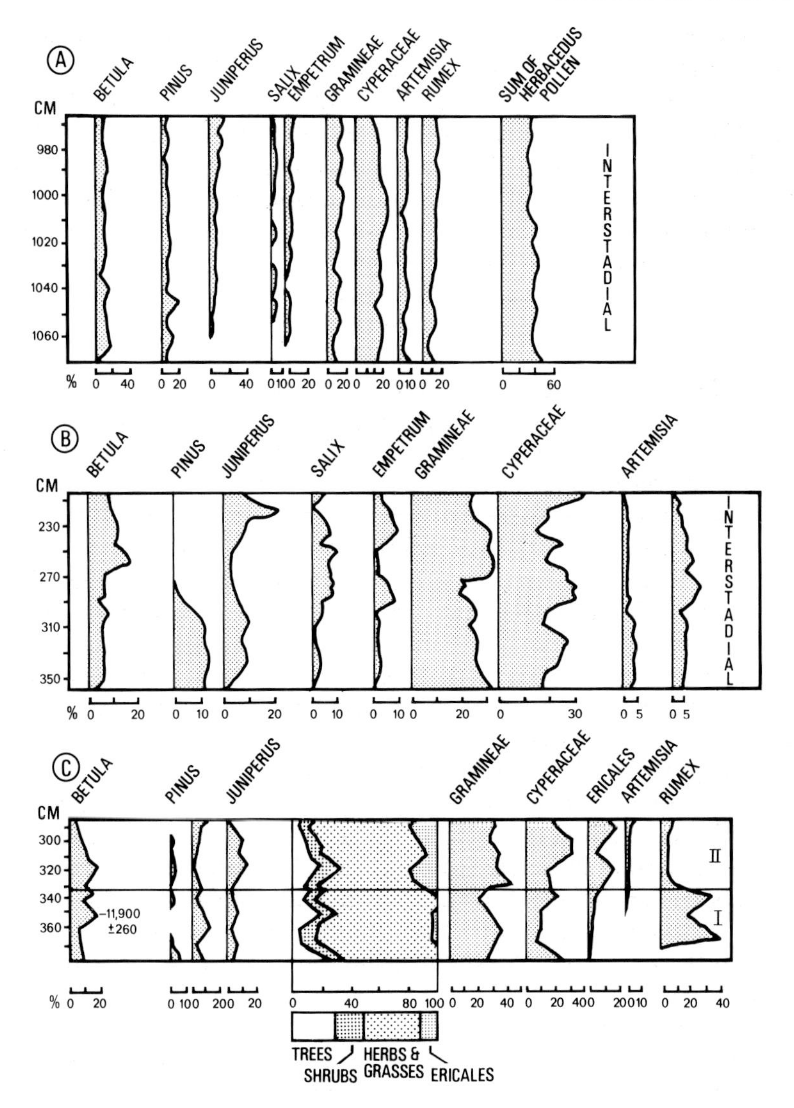

Fig. 4.10. Pollen diagrams (selected taxa) from A. Amulree (Lowe & Walker 1977); B. Blackness (Lowe & Walker 1977); and C. Loch of Park (Vasari & Vasari 1968).

with the generalisation made by Lowe and Walker (1977) and Vasari (1977) and it appears that most of Scotland was characterised by a heath-grassland-dwarf shrub association during the Lateglacial Interstadial. There is however, some debate about the significance of small percentage declines in the representation of woody plants, notably *Empetrum* and *Juniperus* in pollen diagrams from a few sites. It has been suggested that the decline in *Empetrum* and *Juniperus* combined with small lithostratigraphic changes represents a minor climatic recession which has been termed the Older Dryas on the European continent. The Scottish sites in which this 'pollen zone' has been recognised are: Loch of Park, Aberdeenshire (Vasari & Vasari 1968; Vasari 1977); Garral Hill near Keith (Donner 1957); Cam Loch, Sutherland (Pennington 1975b, 1977); Stormont Loch, Blairgowrie (Caseldine 1980). The significance of these sites in relationship to the interpretation of the climatic conditions which prevailed in Scotland during the Lateglacial Interstadial are discussed in the section on climate.

FAUNA

At none of the Lateglacial Interstadial sites shown on Figure 4.6 have fossils of land mammals been recovered. Our knowledge of the mammalian fauna associated with the wastage of the Devensian ice sheet and the development of heath-grass-shrub vegetation is very limited. Although there are numerous records of bones recovered from the glacial and fluvioglacial deposits of Scotland (Lacaille 1954, pp. 79–91), it is often difficult to be certain of their stratigraphic position and especially difficult to be certain that any particular find relates to the Lateglacial Interstadial rather than to the early Flandrian. According to Stuart (1974) the elk (*Alces alces*) reindeer (*Rangifer tarandus*), giant fallow deer (*Megaceros libernicus*) and arctic lemming (*Dicrostonyx torquatus*) did occur in Scotland during the Lateglacial Interstadial but the woolly mammoth (*Mammathus primigenius*) and woolly rhinoceros (*Coelodonta antiquitatis*) did not return.

Perhaps the most interesting fossils yet found in the Scottish Lateglacial deposits have been beetles (*Coleoptera*). Although coleoptera have only been collected and studied at five Lateglacial sites in Scotland (Fig. 4.6) the contribution of such studies to our understanding of the Lateglacial Interstadial environment has been considerable (Coope 1968; Coope 1975; Bishop & Coope 1977). At three of these sites: Roberthill, Redkirk Point and Bigholm Burn, radiocarbon dating of the deposits in which the coleoptera were found indicates that they are representative of the Lateglacial period.

At Roberthill (Bishop & Coope 1977) the coleoptera present indicate that about 13,000 bp this area had a patchy, open vegetation without any trees apart from *Salix* (willow) near pools. Coope (1977) suggests that the climate at that time was very similar to that of the present day. At Redkirk Point (Sample A) specimens of *Bembidion gilvipes* were found and this beetle occurs throughout England and Ireland at present. A second sample (B) taken from Redkirk Point was dated at 11,205 ± 177 bp and it was quite different in terms of its coleopteran fauna from sample A which was at the base of the sequence and therefore older. Sample B was a peat derived from an acid bog with many sedges and mosses and a light cover of *Salix* (willow). All of the relatively 'southern' beetle species recognised in sample A have been replaced by a beetle fauna which has a more northerly range e.g. *Diacheila arctica*. Bishop & Coope (1977, p. 80) state that, "As a group these species would be characteristic of the *Betula* zone in the Scandinavian mountains, that is at an altitude near to the climatic limit for tree growth. At all the sites . . . the deposits laid down between 12,000 and 11,000 bp contain rich insect assemblages characterised by a return of the northern element in the fauna and the disappearance of the thermophilous species." Coope deduces that such a change in the fauna indicates a marked deterioration of climate, at or slightly earlier than 12,000 bp, which continued and culminated with the onset of the Loch Lomond 'glacial/periglacial' environment.

CLIMATE

Any conclusions about the climate of the Lateglacial Interstadial are dependent upon interpretations of lithological, geomorphical, zoological or botanical evidence. In the crudest terms, since the Interstadial refers to a period following the wastage of an ice sheet and preceding the development of an ice cap and small valley glaciers, it may be deduced that the climate changed from 'cold' to 'mild' to 'cold' and that the Interstadial would most likely have a 'mild' climate. The large amount of information now available from pollen analyses indicates a tree-less landscape dominated by grasses, heaths, shrubs, sedges and mosses. It is not surprising then, that many references can be found in the literature to the Lateglacial Interstadial tundra environment of Scotland. Such references would suggest a climate to which the adjective 'mild' or temperate would be inapplicable. Until relatively recently it was assumed that much of the period between 13,000 bp and 11,000 bp could be equated with the Alleröd Interstadial of north-west Europe and that prior to the onset of the Alleröd there had been a three fold climatic oscillation equivalent to the Oldest Dryas-Bölling-Older Dryas of north-west Europe (Gray & Lowe 1977, p. 170).

However, as a result of Coope's work on coleoptera both in Scotland (Bishop & Coope 1977) and in England (Coope 1975) it has been suggested that one major climatic oscillation occurred between 13,000 and 11,000 bp which cannot be subdivided into separate Bölling and Alleröd Interstadials.

Coope's interpretation of the climate of south-west Scotland during the Lateglacial Interstadial suggests that by 13,000 bp the climate had become as warm as at present (Fig. 4.11) but that sometime around 12,000 bp there was considerable climatic deterioration with July average temperatures of 12°C. However, the deep-sea core evidence (Fig. 4.11) indicates that sea surface temperatures to the west of Scotland (Ruddiman *et al* 1977) rose sharply between 14,000 and 13,000 bp and continued to rise until just before 11,000 bp. It seems likely that Scotland experienced mean January temperatures of between −2° and +4°C and mean July temperatures of between 9° and 12°C over the period 13,000 to 11,000 bp. Such a climate would certainly be very marginal for tree growth and this relatively short period of warmer conditions was probably not long enough to allow the migration into, and establishment of trees in Scotland. Such a 'marginal' environment for tree growth would help to explain the absence of significant amounts of tree pollen in the

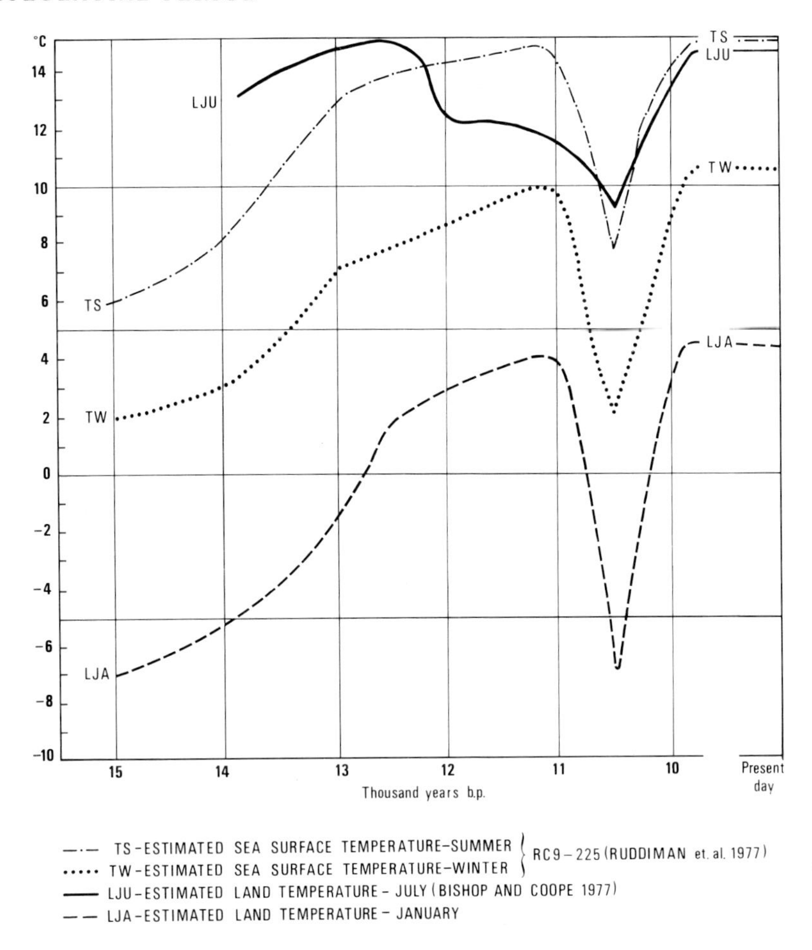

Fig. 4.11. Estimated temperatures: South-west Scotland (Bishop & Coope 1977) and in the NE Atlantic (Ruddiman *et al.* 1977).

pollen diagrams from that period. A major decline in temperatures (Fig. 4.11) about 11,000 bp brought the Lateglacial Interstadial to an end.

The Lateglacial (Loch Lomond) Stadial: circa 11,000–10,000 bp

Despite the problems associated with its chronology there is little doubt that the Loch Lomond Stadial represents a short, sharp deterioration in Scotland's climate. A return to glacial and periglacial conditions is clearly indicated by the geomorphological, lithological and fossil evidence. It is fortuitous that on the one hand this phase of glacial and periglacial conditions, although short lived, was of sufficient duration to be clearly identified in the geological record while, on the other hand, the extent of ice cover was relatively limited so that it did not obliterate all evidence of the previous interstadial conditions. Since the Late Devensian ice sheet completely covered Scotland the ice margins of the Loch Lomond Stadial glaciers are the only maximal ice limits recognisable in Scotland. There is now a very considerable literature relating to the identification of the Loch Lomond Advance (see bibliography in Sissons 1979). J. B. Simpson (1933) clearly identified the limits of these glaciers along the Highland border west of the River Tay and the distinctive landforms and deposits associated with the Lateglacial glaciers which occupied the Loch Lomond basin allowed him to map its piedmont margin and subsequently the other glaciers of this period were named the Loch Lomond Readvance.

The term "readvance" was used by Simpson and many other authors because it was believed that the Devensian ice sheet did not completely waste away before a return to colder conditions allowed the 'dying' ice mass in the higher ground of Scotland to readvance. Although by 1976 (p. 90) Sissons was of the opinion that ". . .total deglaciation by 12,500 radiocarbon years ago seems a conservative suggestion," he still retained the term "Loch Lomond Readvance". In 1970 Sugden argued that in the Western Cairngorms the Loch Lomond Readvance was represented by only minor fluctuations of the margin of the Devensian ice sheet although he has subsequently changed his opinion (Sugden 1980). Peacock (1970) stated that the large volume of ice involved in the Loch Lomond Readvance in western Inverness-shire could not have accumulated in the time available and suggested that active ice existed in the area throughout Lateglacial times. It has now been demonstrated that Scotland was ice free during the Lateglacial Interstadial and it is most unlikely that even in the highest ground did cirque glaciers survive this mild interlude. It is therefore inappropriate to refer to the ice masses which developed during the Lateglacial Stadial as a "readvance".

During the last decade Sissons and several of his research students have undertaken a major research programme to delimit the extent of the Loch Lomond Stadial glaciers (Sissons 1972a, 1973a, b, 1974c, 1975a, 1977a, 1977b, 1977c, 1978, 1979a, 1979b, 1979c, 1980a; Sissons & Grant 1972; Thompson 1972; Gray & Brooks 1972; Sissons, Lowe, Thompson & Walker 1973; Sissons & Sutherland 1973; Gray 1975; Ballantyne & Wain Hobson 1980; Cornish 1981). Throughout this research programme heavy reliance has been placed upon morphological evidence to delimit former ice margins. Sissons (1979a, p. 199) states, "There are scores of end moraines, many of them small, but the longest extend continuously for 15 km or intermittently for 40 km, the highest has outer slopes 30–40 m high, and the broadest individual end moraine ridges are 200 m wide. These clear, often arcuate end moraines record the maximal extent of many of the former glaciers: other clear end moraines often occur immediately behind them. Many glaciers left extensive areas of hummocky moraine . . . and some small ones deposited great quantities of boulders as continuous undulating spreads." While few people would question the significance of morainic ridges as delimiting the former position of ice margins some

Plate 4.2. A. Hummocky moraine—Talla Valley, central Southern Uplands.
B. Hummocky moraine—Upper Glen Truim.

A

B

would argue that such features may not necessarily indicate the maximum extent of the glaciers which produced them and that it is even more dangerous to assume that such maximum limits were necessarily contemporaneous. This assumption is particularly significant in relationship to the attempts which have been made (e.g. Sissons & Sutherland 1976), to reconstruct the mass balance of these glaciers from their supposed ice margins. There is even more controversy about the significance of "hummocky moraine." The distinctive morphology of many valley-floors in both

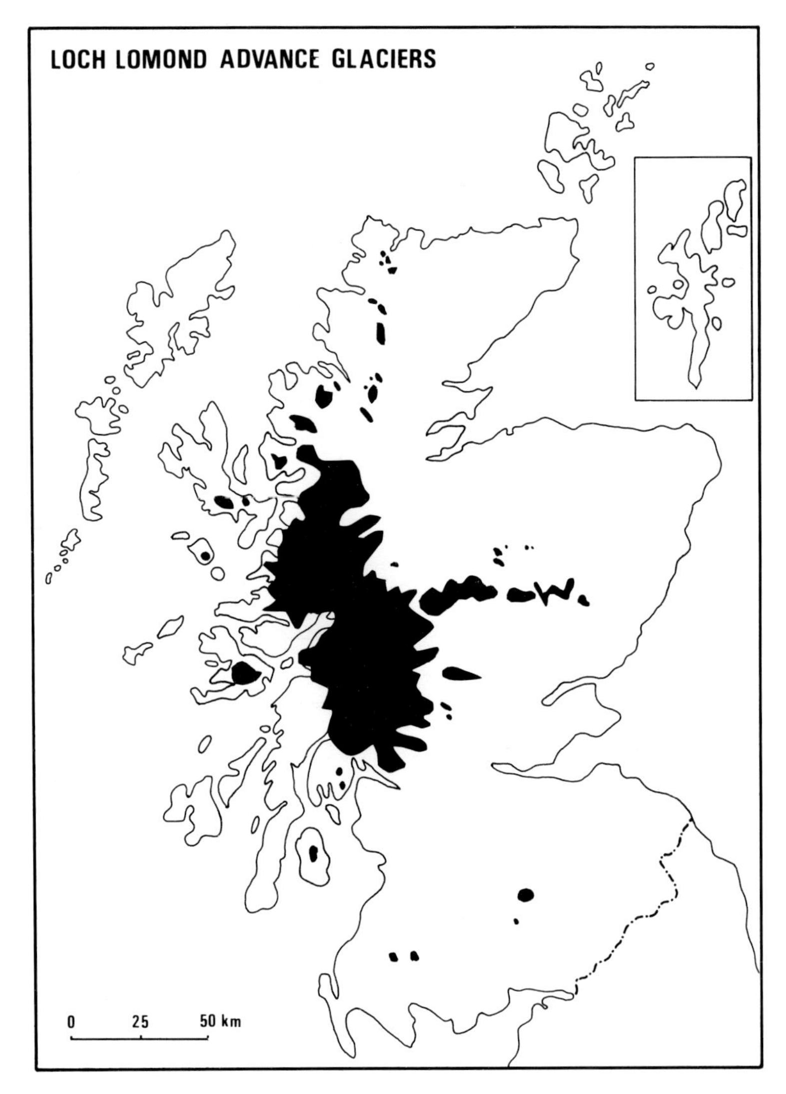

Fig. 4.12.

Loch Lomond advance glaciers.

the Highlands and the Southern Uplands (Pl. 4.2A, B) which were occupied by Loch Lomond Stadial glaciers has been recognised for over one hundred years. The "fresh morainic mounds" have been mapped by many workers (Young 1864; Price 1963; Sugden 1970; Peacock 1970; Sissons 1979; Cornish 1981; May 1981). Steep-sided conical mounds, short ridges aligned in echelon and chaotic assemblages of mounds, ridges and kettle holes are very common. The material constituting these various forms ranges from interbedded silts, sands and gravels through till to large boulders with little or no interstitial material. Little attempt has been made to understand the genesis of these forms yet they are frequently assumed to indicate the former presence of glacier ice and their areal extent has been used to indicate ice margins. Clapperton, Gunson & Sugden (1975, p. 710) state, ". . .the identification of the limits of the Loch Lomond Readvance by mapping the distribution of 'fresh hummocky moraine' must now be considered to be a very precarious technique." Until the question of the origin and significance of 'hummocky moraine' has been resolved a major uncertainty will remain about the ice limits which have been reconstructed for some of the Loch Lomond Stadial glaciers.

Radiocarbon dates directly related to the glaciers of the Loch Lomond Stadial have been obtained from five sites. Ice-transported shells from the Menteith moraine (Sissons 1967a) in the Forth Valley have been dated as 11,800 ± 170 bp, shells from fluvioglacial deposits inside the Loch Lomond moraine (Sissons 1967c) gave a date of 11,700 ± 170 bp, and shells from the Kinlochspelve moraine on Mull provided (Gray & Brooks 1972) a date of 11,330 ± 170 bp; shells from glacially disturbed marine clay beneath deposits of the Loch Creran glacier (Peacock 1971) yielded dates of 11,430 ± 220 bp, 11,530 ± 210 bp and 11,805 ± 180 bp; shells in the Gare Loch end-moraine (Rose 1980) provided a date of 11,520 ± 250 bp. These dates indicate that the glaciers of the Loch Lomond Stadial reached low ground (in some instances sea-level) sometime after 11,500 bp.

LOCH LOMOND ADVANCE: GLACIERS, LANDFORMS AND DEPOSITS

Figure 4.12 shows the maximum extent of glacier ice in Scotland during the Loch Lomond Stadial. At this scale it is not possible to indicate many of the small individual cirque glaciers which developed or the extensive 'nunatak' areas which stood above the west Highland glacier complex. The major area of glacier development was in the high ground of the Western Highlands stretching from Loch Alsh in the north-west to Loch Lomond in the south-east, a distance of some 150 km. Smaller ice masses developed in the north-west Highlands, south-eastern Grampians, the Cairngorms, the inner Hebridean islands of Skye, Rhum, Mull and Jura, on Arran and to a very limited extent in the Southern Uplands. This distribution pattern has a distinctly 'western' emphasis and there is of course strong correlation with areas of high ground.

THE WEST HIGHLAND GLACIER COMPLEX

It is not surprising that the backbone of the Scottish Higlands stretching from Loch Alsh to Loch Lomond was the main area for glacier development during the Loch Lomond Stadial. Although the area is deeply dissected by glacial troughs produced during previous glaciations there are numerous areas of high ground (over 800 m) plus hundreds of corries suitable for snow accumulation. The term glacier complex has been used because it is most likely that the short period of the Lateglacial Stadial was insufficient to allow an extensive ice-cap to develop except perhaps over the Rannoch Basin (Fig. 4.13). Valley glaciers mainly fed from corries

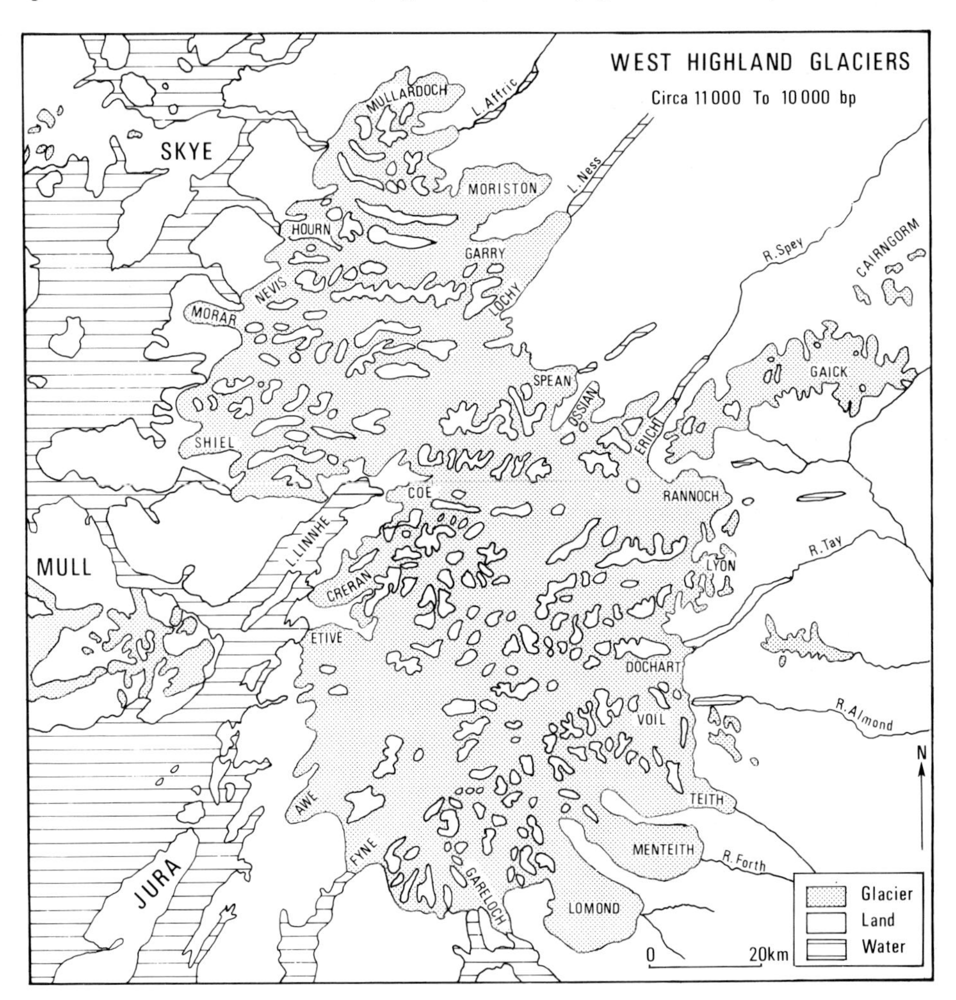

Fig. 4.13. West Highland glacier complex.

or from plateau glaciers dominated this area. There were four main areas of ice accumulation: the area to the north-west of the Great Glen; the mountains between

Glen Spean and Glen Etive; the Rannoch Basin; the mountains between Glen Dochart and Loch Long. The longest individual ice streams, each about 50 km in length, occupied the troughs of Loch Lomond, Loch Awe, Loch Rannoch and Loch Garry. It is interesting that one of the largest glaciers in the system occurred in the south and occupied the Loch Lomond basin (Fig. 4.14) producing a large piedmont lobe some 20 km wide. The total volume of ice in the Loch Lomond glacier was in excess of 80 km^3. Around the frontal margin of this glacier is an 'end-moraine' complex consisting of morainic mounds and ridges (Pl. 4.3), fluvioglacial mounds and ridges, deltas and eskers which extends for over 40 km and ranges in altitude from near sea level to 300 m. The relationship of this end-moraine complex to the Loch Lomond valley to the north, suggests that in the rock basin some 25 km upstream from the ice margin, the glacier was 600 m thick.

The Loch Lomond glacier dammed up lakes in the Blane, Endrick and Fruin Valleys (Figs. 4.15, 4.16). An ice-dammed lake in the Drymen area was postulated by J. Geikie in 1894. With the Loch Lomond basin filled by glacier ice the valleys of the River Endrick and its tributary the Blane Water, would have been dammed to produce a lake some 14 km long in the Endrick Valley and 10 km long in the Blane Valley. The former existence of this lake is proved by the occurrence of extensive lake-bottom deposits in both valleys, which are known from boreholes to be over

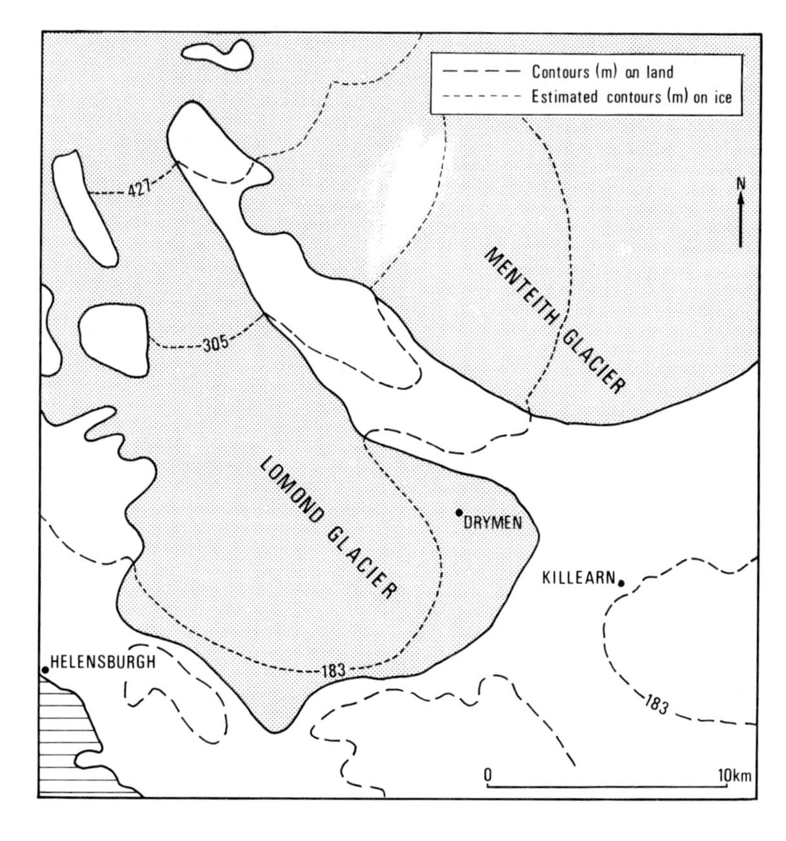

Fig. 4.14.

The Loch Lomond glacier.

12 m thick. There are also massive deltaic accumulations of sands and gravels in the Drymen area and 3 km east of Fintry. There are also some shoreline 'benches' developed on the flanks of some of the drumlins which occupy the floor of the Endrick Valley. The level attained by the lake was controlled by the position of the ice margins in the Forth Valley and in the Loch Lomond basin and by the level of the 'overflow' meltwater channels east of Strathblane and along the watershed between the Endrick Valley and the Forth Valley (Fig. 4.15). On most published maps the maximum extent of the Loch Lomond glacier is as shown on Figure 4.15 and therefore 'Lake Endrick' and 'Lake Blane' must have been linked and the maximum level of this water body during the Loch Lomond Advance was determined by the channel east of Strathblane at an approximate altitude of 80 m. As the lake level fell below this altitude the channel to the north-east of Drymen began to function allowing water to penetrate the Forth glacier. This explanation does not

Plate 4.3. Moraine ridge (2–3 m high) marking the western limit of the Loch Lomond Glacier in Glen Fruin.

account for the occurrence of at least 8 m of lake clays which occur on the Culcreuch estate near Fintry at an approximate altitude of 90 m and a large deltaic sequence 3 km east of Fintry the upper surface of which attains an altitude of approximately 150 m. Either, these two deposits relate to the main Late Devensian phase of ice sheet deglaciation and represent a period of lake development when 'Lake Endrick' was not linked to 'Lake Blane' and therefore its level was not controlled by the channel to the east of Strathblane, or the maximum extent of the Loch Lomond ice margin shown on Figure 4.15 is incorrect. If the Loch Lomond ice

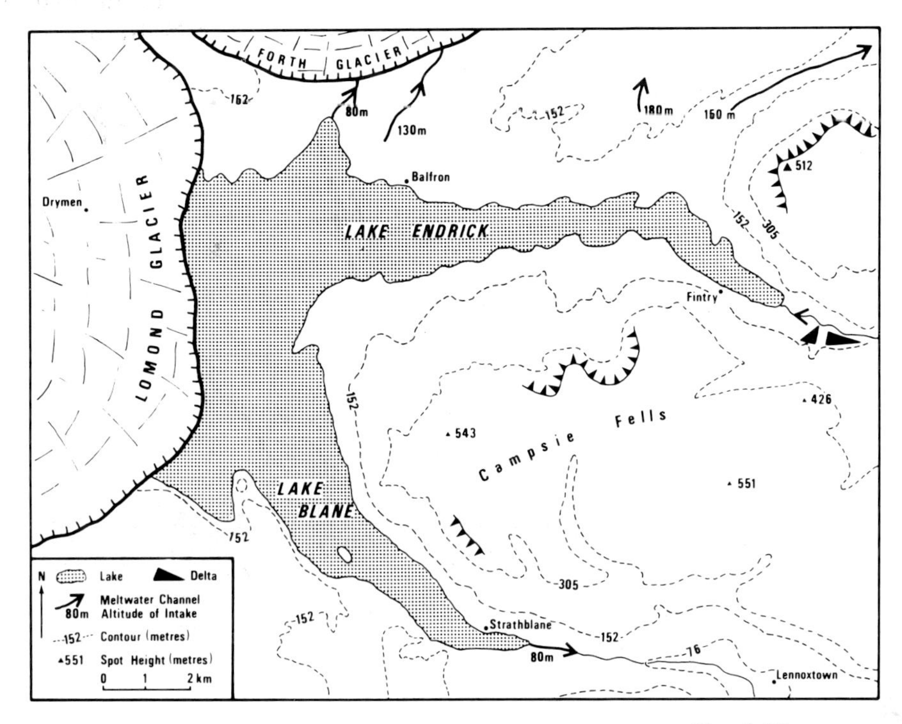

Fig. 4.15.

Proglacial Lakes: Endrick and Blane.

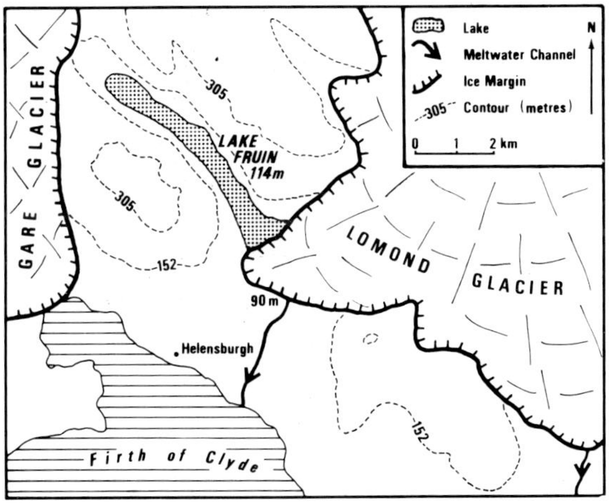

Fig. 4.16.

Proglacial Lake Fruin.

margin in fact extended eastwards some 4 km it would have sealed off the Blane Valley and the lake in the Endrick Valley would have been a separate water body, the level of which could have risen to an altitude of approximately 150 m. The outlet for such a lake would have initially been the channel 4 km due north of Fintry followed subsequently by the channels to the north and west of Balfron.

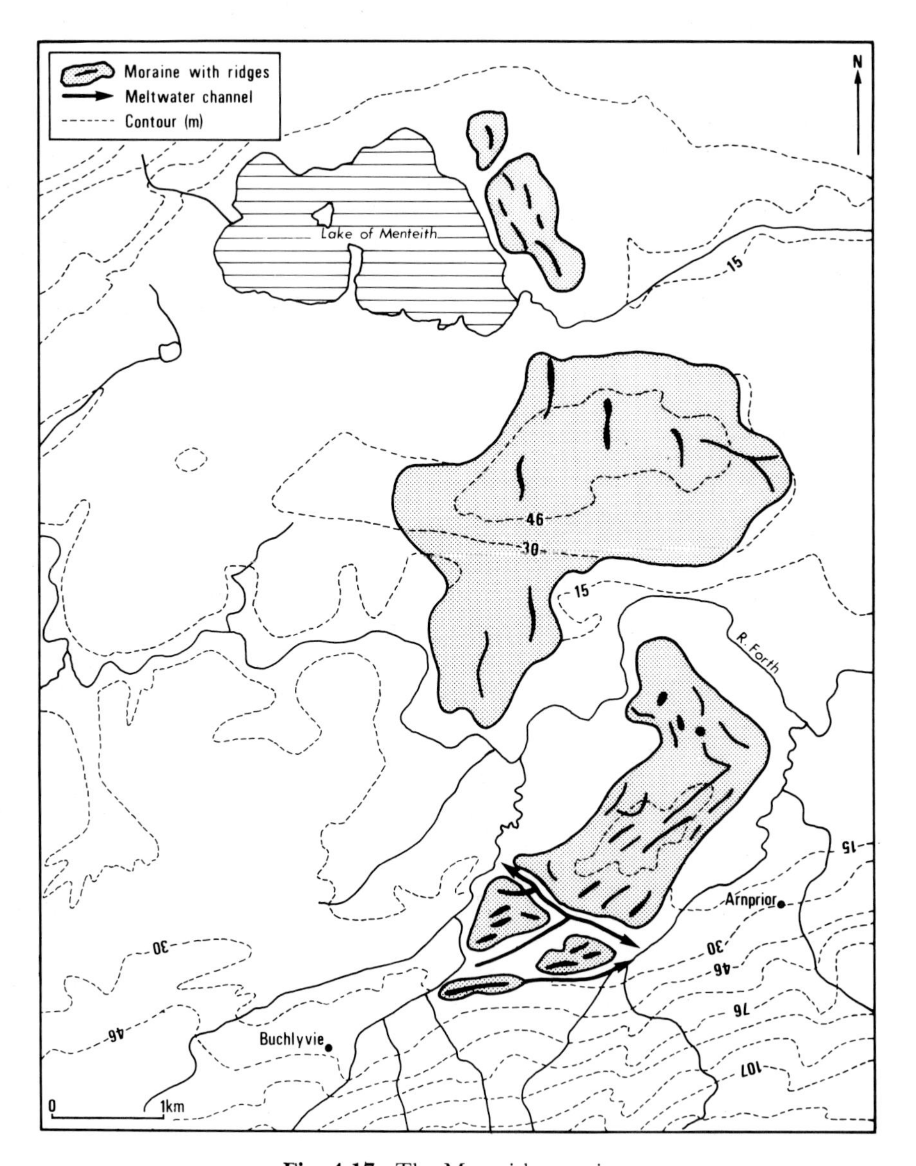

Fig. 4.17. The Menteith moraine.

At the time of the maximum extent of the Loch Lomond glacier a lake about 6 km long also developed in the Fruin Valley on the western margin of the Lomond glacier. Boreholes through the floor of that valley have penetrated some 10 m of sands and fine gravels resting on more than 12 m of fine silt. The age of these sediments is unknown but the western margin of the Lomond glacier would only have had to reach an altitude of about 115 m on the north facing slopes of Glen Fruin to allow a lake to develop. Once ice retreat had taken place to allow the channel north-east of Helensburgh to function (its intake is at approximately 90 m) the lake in Glen Fruin would have drained.

In the upper Forth Valley there is another arcuate morainic complex some 20 km long associated with a piedmont ice lobe (Fig. 4.17). The moraine consists of ridges and mounds of sand and gravel and glacier transported marine clay with shells (Gray & Brooks 1972). Simpson (1933) described a section in the moraine on the banks of the Lake of Menteith where 3 m of dark grey clay containing fragments of *Mytilus edulis* are overlain by about 9 m of sand and gravel. It was from this section that the shells were collected which produced a radiocarbon date of 11,800 ± 170 bp. Shell assemblages from this site indicate boreal to sub-arctic water conditions and therefore that sea temperatures were lower during the late stage of the Lateglacial Interstadial than they are to-day.

The eastern limit of the West Highland glacier complex (Thompson 1972; Smith *et al* 1978) follows a line northwards from Callander through Lochearnhead and Killin to Glen Lyon (Fig. 4.13). This south-east segment of the glacier system was largely nourished by numerous local valley glaciers and it is this circumstance that perhaps best explains the relatively limited eastward extension of these glaciers.

In contrast to the south-eastern part of the glacier system, glaciers occupying the Rannoch, Ericht, Ossian and Treig Valleys were mainly nourished by ice originating on the Moor of Rannoch. The presence of lateral moraines has enabled Sissons (1976a) to calculate that the ice surface sloped eastwards at an average gradient of 26 m/km for 10 km and then at 55 m/km for 3 km to its terminus beyond which, at the eastern end of Loch Rannoch, an outwash plain filled the valley floor.

Sissons (1976a) also states that he believes the ice on Rannoch Moor was more than 400 m thick and that the ice shed altitude was 850–900 m. Both of these figures would suggest that many of the mountain ridges which have altitudes between 600 and 900 m would have stood above the valley glaciers. Although for a short period it is likely that a true ice cap may have existed over Rannoch Moor with the complete burial of the land in that area by ice, the remainder of the West Highland glacier complex was characterised primarily by valley glaciers.

Ice flowing northwards from the Ben Nevis Range and Rannoch Moor, down Glen Ossian and the valley now occupied by Loch Treig joined with ice moving east across the Great Glen (Peacock 1970) from Glen Loy and Glen Arkaig to produce a major ice mass in Glen Spean which in turn resulted in the development of Scotland's most famous Quaternary features: the Parallel Roads of Glen Roy (Pl. 4A, B). The movement of glacier ice into Glen Spean from the west and south produced an ice dam at the southern ends of Glen Roy, Glen Gloy and the eastern

A. Lithographic sketch by Mr. Albert Way of the Parallel Roads of Glen Roy, originally printed in a paper by Charles Darwin in the *Phil. Trans. Roy. Soc.* in 1834.

Plate 4.4.

B. The Parallel Roads of Glen Roy.

end of Glen Spean which allowed the development of a series of ice-dammed proglacial lakes (Fig. 4.18) and their associated shore-lines (Parallel Roads).

The Parallel Roads of Glen Roy have received a great deal of attention since Jamieson produced his classic paper in 1863—over 35 publications relating to these

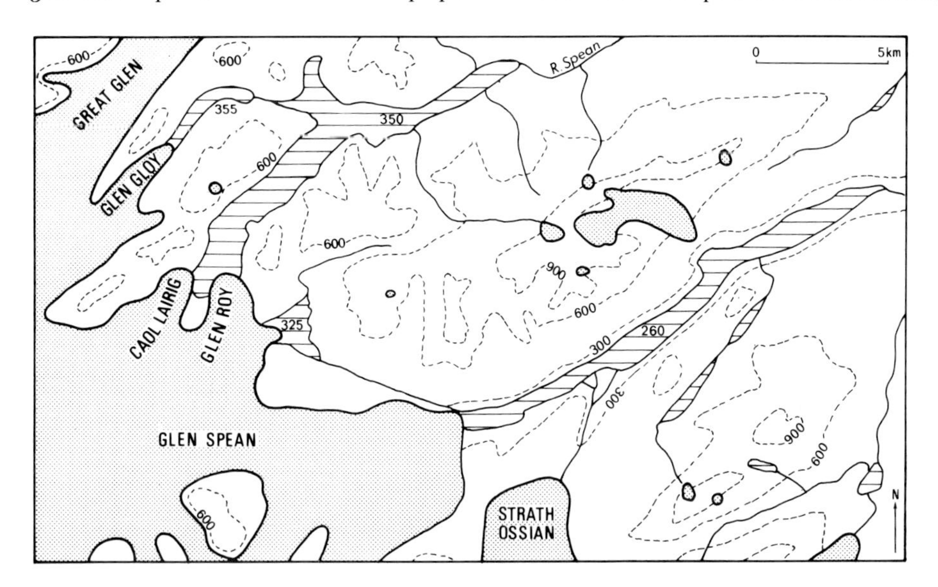

Fig. 4.18A.

Ice-dammed lakes in Glen Roy, Glen Gloy and Glen Spean (Sissons 1979).

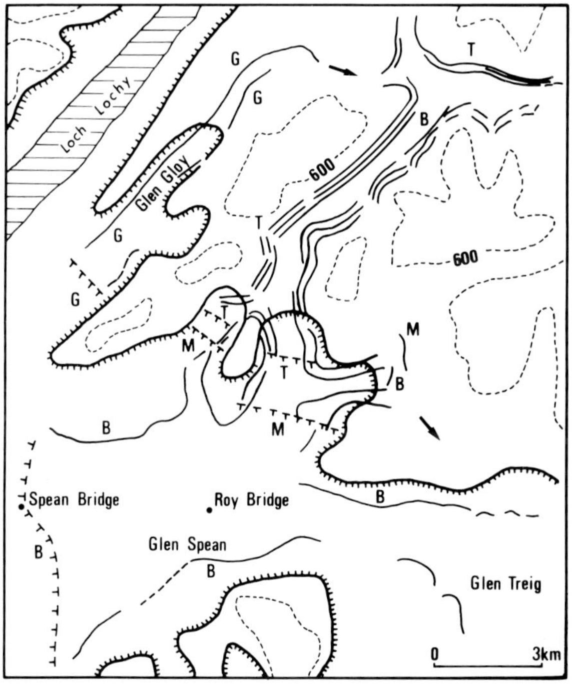

Fig. 4.18B.

Lake shorelines (Parallel Roads) in Glen Roy, Glen Gloy and Glen Spean. Letters T, M, B and G identify respectively Top, Middle, Bottom and Gloy Roads and final positions of the corresponding ice dams. (Sissons 1978.)

features had been produced by 1885. The parallel roads were variously attributed to a great flood, deposition along a glacier margin, marine shorelines, shorelines of debris-dammed or ice-dammed lakes. When Agassiz first brought the theory of glaciation to Britain in 1840 he suggested that the parallel roads of Glen Roy were the shorelines of ice-dammed lakes and it is this explanation which has been most widely accepted.

Although the subject of much general attention, the Glen Roy area had had no serious scientific investigation of its glacial phenomena for more than 70 years, until J. B. Sissons re-examined the area and published his conclusions in a series of papers (Sissons 1978, 1979c, 1979d). Some of the main problems associated with the lake-shoreline hypothesis, namely their occurrence on steep slopes in situations in which only a limited fetch could have generated wave activity, prompted Sissons to re-examine the evidence relating both to the ice-dams and the shorelines. There are three shorelines in Glen Roy, four in Caol Lairig, one in Glen Gloy and one in Glen Spean (Fig. 4.18) and occasional traces of shorelines at other levels.

Jamieson (1863) believed the shorelines were mainly the result of deposition but in 1892 he mentioned shorelines eroded in rock near the head of Glen Roy. He also stated that the 'roads' are broader and have a greater vertical extent where wave action was most effective owing to exposure to west or south-west winds.

Sissons measured the characteristics of the shorelines at 787 points (Sissons 1978). He points out that the 'roads' were normally formed by erosion at the back and deposition at the front and that they are normally cut in rock rather than glacial drift. Although the most common cause of local increases in road width was debris supplied by the principal streams entering the lakes, road width is clearly related to aspect, with roads facing south, south-west and west being especially favoured. Sissons (1978, p. 242) states that, ". . .on the east side of Glen Roy (304868) . . . the top road is cut clean across metamorphic rocks over a width of 7 m. . . . In upper Glen Roy between the Roy and the Burn of Agi . . . the middle road is a rock platform up to 12 m wide, with an irregular surface, the bare rock cliff at the back being up to 5 m high and sloping at angles up to 70°." Such platforms cut in solid rock are hard to explain in terms of wave action in a relatively short-lived lake with limited fetch. Sissons (1978, p. 243) points out that, "The typical angular debris of the roads is consistent with waves of limited power and hence implies they cannot have eroded into firm bedrock."

Sissons concludes that the shorelines were produced by powerful frost action in the critical environment along the shore at each lake level. Such frost action could not have transported large slabs of rock away from the shorelines and wave action would not remove them either. Sissons suggests that the debris was incorporated in the ice that would have formed on the lake surface each winter, subsequently being carried away when the ice broke up in the spring. This suggestion of ice-floe transport of debris is supported by the presence of angular rock fragments within the fine grained sediments on the floors of the former lake sites.

Sissons has applied his theory of rapid shoreline patform development during the Lateglacial Stadial in south-east Scotland (Sissons 1976c) and in the Firth of Lorn (Sissons 1974d) to the shorelines in Glen Roy, Glen Gloy and Glen Spean.

The severity of the climate during the Lateglacial Interstadial has been established by various types of evidence and it seems reasonable to explain the development of the lake-shorelines in terms of wave action in an intense periglacial environment in which lake ice played an important part.

The former presence of ice-dammed lakes in Glen Roy, Glen Gloy and Glen Spean is known on the basis of evidence other than the presence of shorelines. Bottom-deposits consisting of laminated silts and clays often overlie, or alternate with, beds of sand and gravel on the floors of the former-lake sites. There are also some large deltas which were built out into the ice-dammed lakes. One such delta divides Loch Laggan into two parts. This delta was built up from outwash from the Strath Ossian glacier while several flat-topped features represent the remains of a delta associated with the Treig glacier (Fig. 4.19).

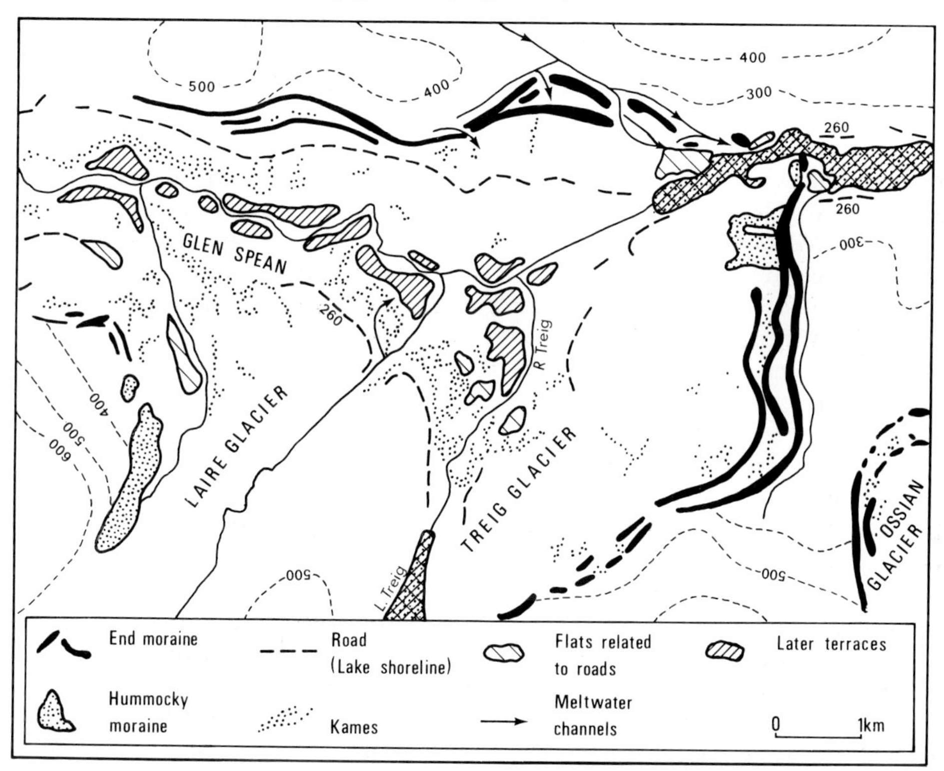

Fig. 4.19. The glacial and fluvioglacial landforms of Glen Spean and Glen Treig (Sissons 1977d).

Deltas are absent from Glen Gloy but are common in upper Glen Roy, in Glen Turret and in Caol Lairig. A very large delta, related to the lowest shoreline (260 m) occupies the ground between the River Roy and River Turret and ends in steep bluffs cut by these rivers. Further down Glen Roy, on its south-east side, two other streams built out delta-fans whose apexes are at the altitude of the lowest shoreline.

The highest shorelines in each glen record the maximum extent of the ice-dammed lakes (Fig. 4.18). The Glen Gloy lake at 355 m overflowed at its north-eastern end into the Spey Valley. At the same time, a small independent lake at 325 m existed to the east of lower Glen Roy and its waters overflowed southeastwards to run along the ice margin in Glen Spean to join the ice-dammed lake in the Spean Valley which reached an altitude of 260 m. The level of this lake was related to a col at the eastern end of Loch Laggan which carried overflow waters into the Spey drainage.

The ice front retreated for some 5 km down Glen Gloy without affecting the level of the ice-dammed lake in that valley. However, it only required a retreat of the Glen Roy ice front of about 1 km to make the 325 m col available and thus lowering the level of the Glen Roy Lake by 25 m.

The dimensions of these ice-dammed lakes were considerable. At their greatest extent the Glen Roy Lake was 16 km long and had a maximum depth of 200 m; the Glen Gloy Lake was 9 km long with a maximum depth of 200 m; the Glen Spean Lake was 35 km long, 6 km wide at the ice dam (near the site of Spean Bridge) and 170 m deep, and at this stage this lake had an extension up Glen Roy for 13 km producing a shoreline at 260 m.

The final drainage of these lakes was almost certainly through subglacial tunnels. The Glen Gloy Lake drained northwards via a subglacial gorge into Loch Lochy in the Great Glen. The Spean Lake probably drained catastrophically westwards beneath ice and utilised the gorge west of Spean Bridge.

Ice originating in the high ground to the west of the Great Glen moved down Glen Arkaig, Glen Lochy and Glen Garry to fill the Great Glen with ice as far north as Fort Augustus. A major glacier system also moved down Glen Loyne and Glen

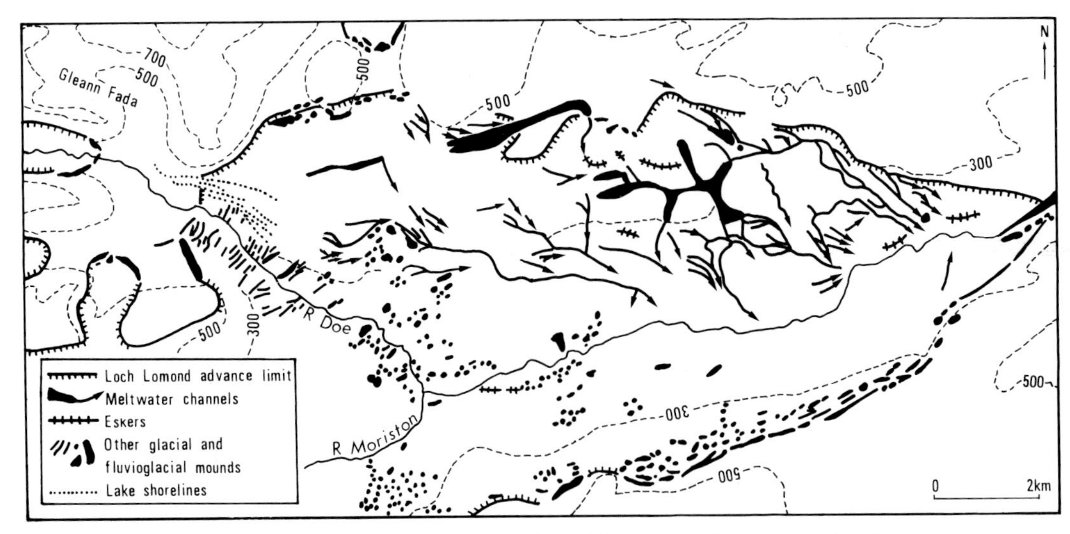

Fig. 4.20. Glacial and fluvioglacial landforms associated with the Loch Lomond advance in Glen Moriston and Glen Doe (Sissons 1977c).

Moriston to produce a mass of ice 8 km wide in lower Glen Moriston. The limits of this ice mass (Fig. 4.20) have been clearly identified by Sissons (1977c) and in Glen Doe, on the northern side of Glen Moriston, Sissons mapped seven shorelines, each two to twenty metres wide, indicating the former presence of an ice-dammed lake, the surface of which ranged between 405 and 304 m in altitude. These shorelines are similar to those described in Glen Roy in that they are partly the result of erosion accompanied by deposition along the front of the features. The former presence of the Glen Moriston ice-dammed lake is also demonstrated by the occurrence of lake floor deposits, overflow channels and marginal cross-valley moraines. Sissons suggests that the lake was drained intermittently by 'jokulhlaups'—catastrophic drainage along and beneath the ice margin, in open ice-walled channels and subglacially. The shorelines of the ice-dammed lake in Glen Moriston leads Sissons (1977c, p. 224) to suggest "...ice-dammed lakes may have been more common during the deglaciation of British Uplands than is currently believed."

One other location at which lake-shorelines have been recognised is Loch Tulla on the south-east side of Rannoch Moor in Argyll. They are different from the lake-shorelines described in the Glen Roy area and in Glen Moriston because they are not the product of ice dams at the maximum extent of the Loch Lomond Advance. The Loch Tulla shorelines occur very close to the centre of ice dispersal and were developed at a very late phase of deglaciation. They were first recognised by Milne (1847), who believed they were produced by a lake, dammed up by a debris barrier in Glen Orchy. Chambers (1848) thought they were shorelines of an ancient sea. Mathieson & Bailey (1925), and Gregory (1926), independently proposed that a series of ice-dammed lakes had occupied the Tulla Basin during the final stages of the last glaciation.

Recent work by Ballantyne (1979), indicates that the Loch Tulla shorelines are from two to twenty metres wide and are predominantly erosional features cut in drift and that the shorelines are absent in the western part of the Tulla Basin (Fig. 4.21). Ballantyne suggests that the shorelines were formed in a lake produced

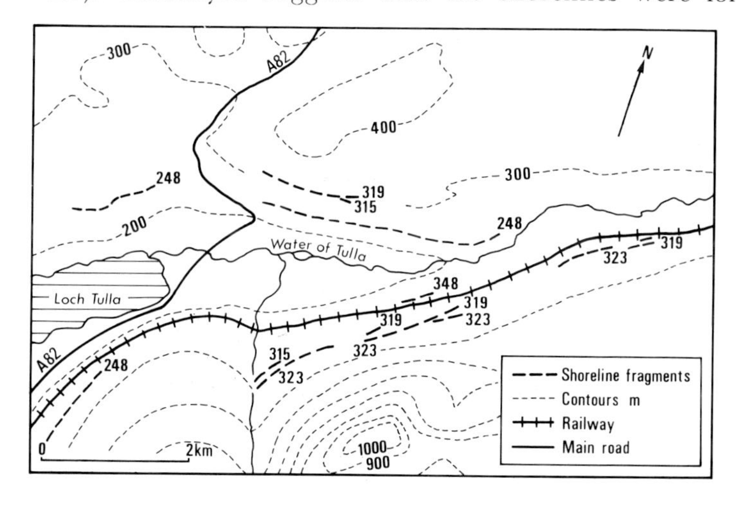

Fig. 4.21.

Lake shorelines in the Tulla Basin (Ballantyne 1979).

during downwastage of the ice mass accompanied by a general westward retreat of the ice margin. The lake originally drained over cols towards the east and north but final drainage was through the ice dam to an outlet at Lochan na Turaiche, 9 km to the west.

To the north-west of the Great Glen the ice-shed was located well to the west and therefore only relatively short valley glaciers (5–10 km long) descended to the sea lochs such as Loch Duich, Loch Hourn and Loch Nevis (Fig. 4.13). Since relative sea-level was probably about 10 m higher in these areas during the Lateglacial Stadial than at present, the limited westward penetration of these valley glaciers into the sea lochs may well be explained by the rapid calving of icebergs in the marine environment during the relatively short period during which the glacier fronts were at sea-level. Such a view is supported by the greater extension of the Morar and Sheil glaciers which most probably were not affected by marine environments.

A similar set of circumstances may explain the surprisingly limited extension of the glaciers in Loch Linnhe and Loch Leven. The termination of both these glaciers is marked by massive accumulations of outwash gravels at Corron Ferry and Ballachulish respectively (Fig. 4.22). One might have expected these glaciers which were fed by large, high altitude, accumulation areas to have extended further down

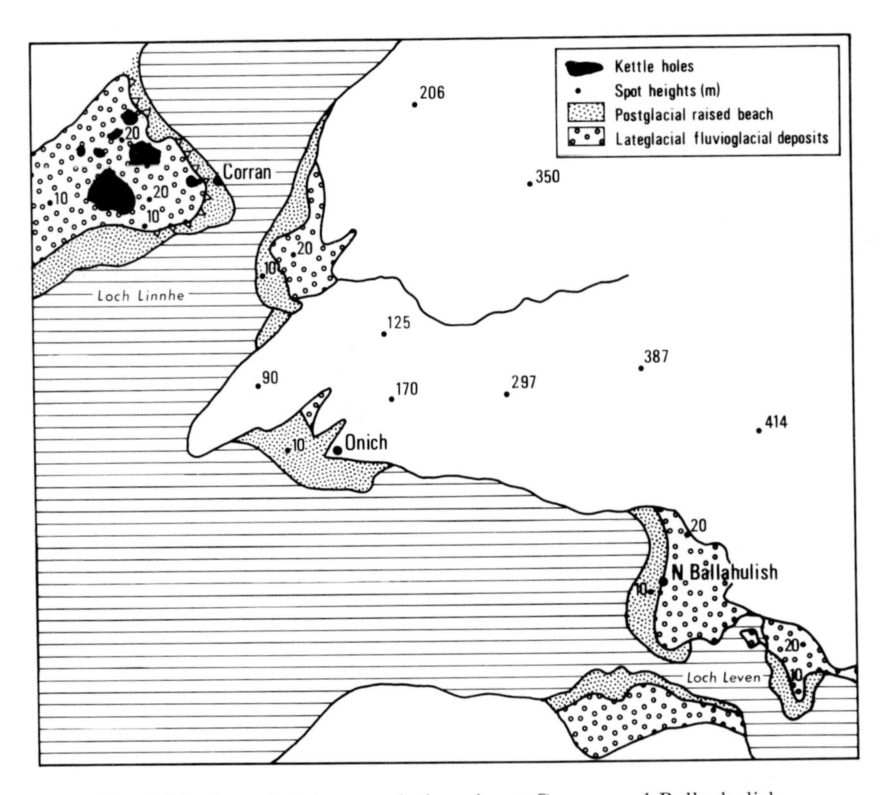

Fig. 4.22. Lateglacial outwash deposits at Corran and Ballachulish.

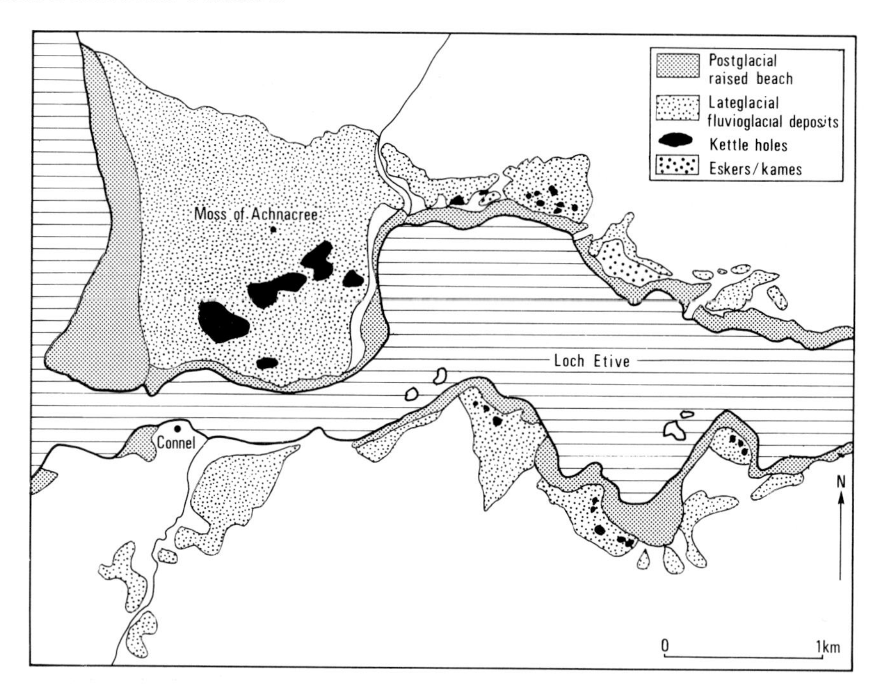

Fig. 4.23. Lateglacial outwash deposits at Loch Etive (Gray 1975).

Loch Linnhe but again the deep water marine environment may have inhibited their extension.

Apart from the two ice lobes which occupied Loch Creran and Loch Etive (Fig. 4.23), the ice marginal landforms of these two glaciers having been mapped in detail (McCann 1961, 1966a; Peacock 1971; Gray 1975), the extent of the valley glaciers along the south-western margin of the West Highland Glacier complex (Fig. 4.13) is only poorly known. No detailed studies have been published relating to the former glaciers in Lochs Awe, Fyne and Long but if the ice margins indicated in these valleys on Figure 4.13 are correct then at least the Loch Awe and Loch Fyne glaciers must owe their large size (40 to 50 km in length) to the extensive accumulation areas around the southern edge of the Moor of Rannoch.

In the Gare Loch, opening into the Clyde estuary, the limit of the Loch Lomond Advance glacier is clearly defined by an 'end-moraine' and lateral moraines (Rose 1980). These features were described by McCallien (1937) and related, on the basis of shoreline distributions to the valley glaciation of the Loch Lomond Advance by Anderson (1949). The age of the features has been confirmed by the occurrence within the morainic ridge of till derived from Clyde Beds, with shells which yielded a date of 11,520 ± 250 bp (Rose 1980).

The West Highland glacier complex can probably be used as an analogue of the early stages of ice sheet development in Scotland. Even in the short period of time during which environmental conditions deteriorated sufficiently for glaciers to develop, probably less than 1,000 years, major ice streams descended from the high ground in the west and north, down to sea level. If climatic conditions had remained severe for another thousand years then a major ice sheet would have developed as the valley glaciers over-topped their divides to coalesce and form an ice dome capable of supplying ice streams to fill the central Lowlands and the Firth of Clyde as happened during the main Devensian glaciation.

THE SOUTH-EAST GRAMPIANS

The high ground south of Loch Earn contained several small valley glaciers (Fig. 4.13) less than 5 km long which descended to altitudes of about 300 m. A

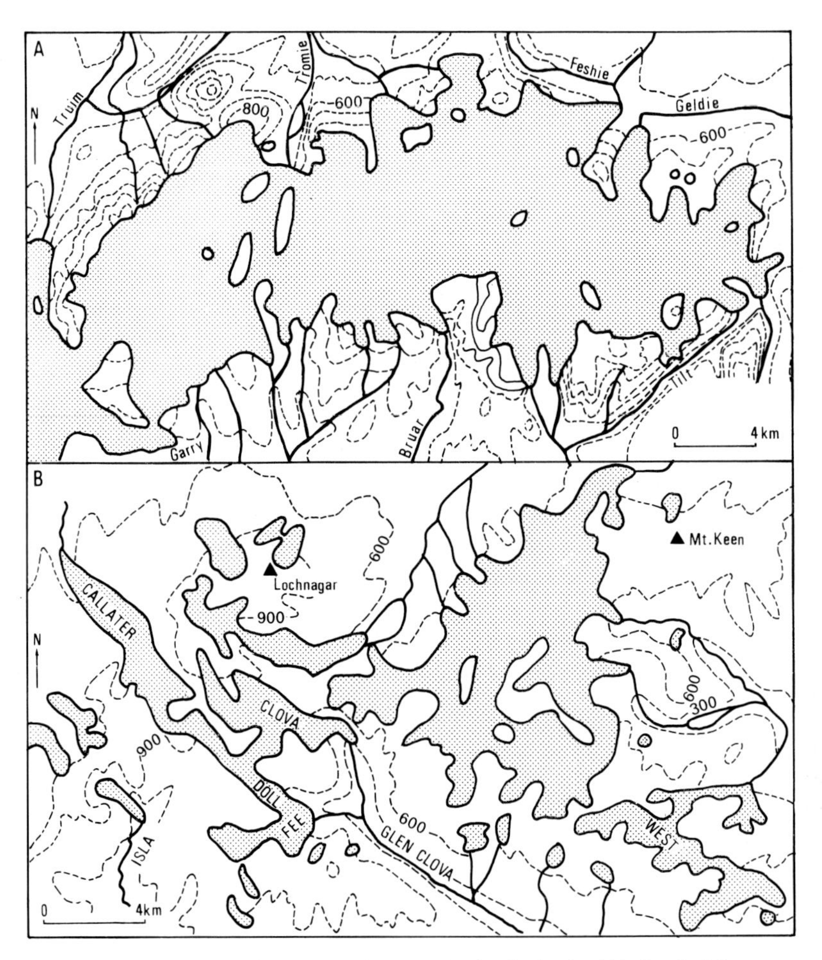

Fig. 4.24. Loch Lomond Advance glacier limits in (A) the Gaick area and (B) the Lochnagar–Glen Clova area of the Eastern Grampians (Sissons 1974 and 1976).

glacier 20 km long occupied Glen Almond to the south-east of Loch Tay. However, it is in the high ground south of the Dee Valley and extending from Drumochter Pass in the west to Mount Keen in the east (Fig. 4.24) that detailed investigations by Sissons & Grant (1972), Sissons (1974c) and Sissons & Sutherland (1976) have revealed the limits of ice caps and valley glaciers which developed during the Lateglacial Stadial. On the Gaick Plateau at an altitude of between 700 and 930 m a plateau ice cap developed over an area of 360 km^2 with an average thickness of 110 m. The surface of the ice cap reached altitudes of between 800 and 900 m with occasional nunataks. The ice descended from the plateau to about 600 m on the northern side and to between 400 and 500 m on the southern side. The limit of the Gaick ice cap is only marked by end moraines for less than 15 km of the total ice marginal length of 180 km. Sissons (1974) used the distribution of meltwater channels and hummocky moraine to differentiate between areas affected by the Lateglacial ice cap and those only covered by the main Devensian ice sheet. On this basis, Sissons identified the margins of the ice cap and its associated outlet glaciers.

The extent of ice cover in the Lochnagar-Mount Keen area was much smaller (65 km^2) than on the Gaick Plateau. Major valley glaciers (4–8 km long) occupied Glen Callater, Glen Doll, Upper Glen Clova, and Glen Muick while an ice cap covered the plateau to the north of Glen Clova. Both the Glen Callater and Loch Muick valley glaciers descended to about 450 m while the Glen Clova and Glen Doll Glaciers descended to about 300 m.

THE CAIRNGORM MOUNTAINS

There has been considerable debate about the extent of glacier ice in the Cairngorms during the Lateglacial Stadial (Sugden 1970; Sugden & Clapperton

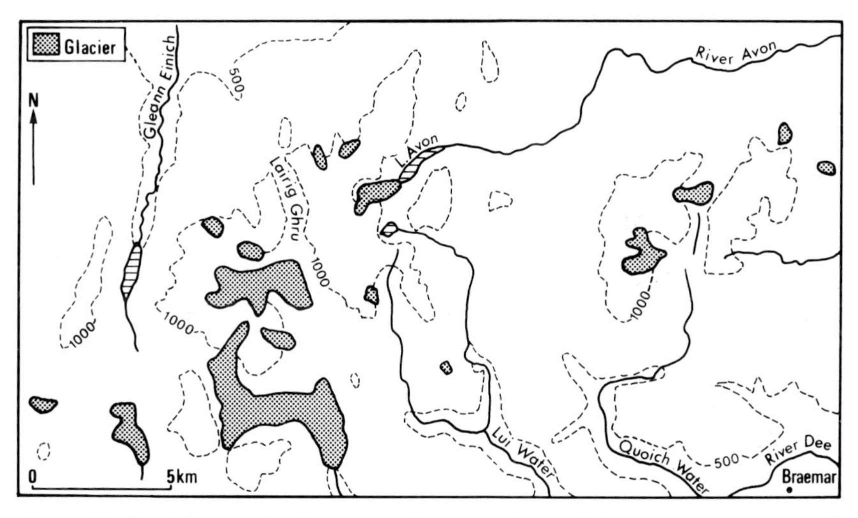

Fig. 4.25. Loch Lomond Advance glacier limits in the Cairngorms (Sissons 1979).

1975; Clapperton, Gunson & Sugden 1975; Sissons 1979b; Sugden 1980). Much of this debate stems from the lack of radiocarbon dates for any of the ice marginal positions which have been postulated along with a lack of knowledge about the genesis and significance of hummocky moraine. These are problems which are not limited to the Cairngorm area alone. However, Sugden (1980) points out that he does agree with the interpretation of the postulated Loch Lomond Advance limits suggested by Sissons (1979b) and it is these limits which are discussed below.

Sissons (1979b) has established the limits of seventeen glaciers that developed in the Cairngorms during the Loch Lomond Stadial (Fig. 4.25). The largest glacier

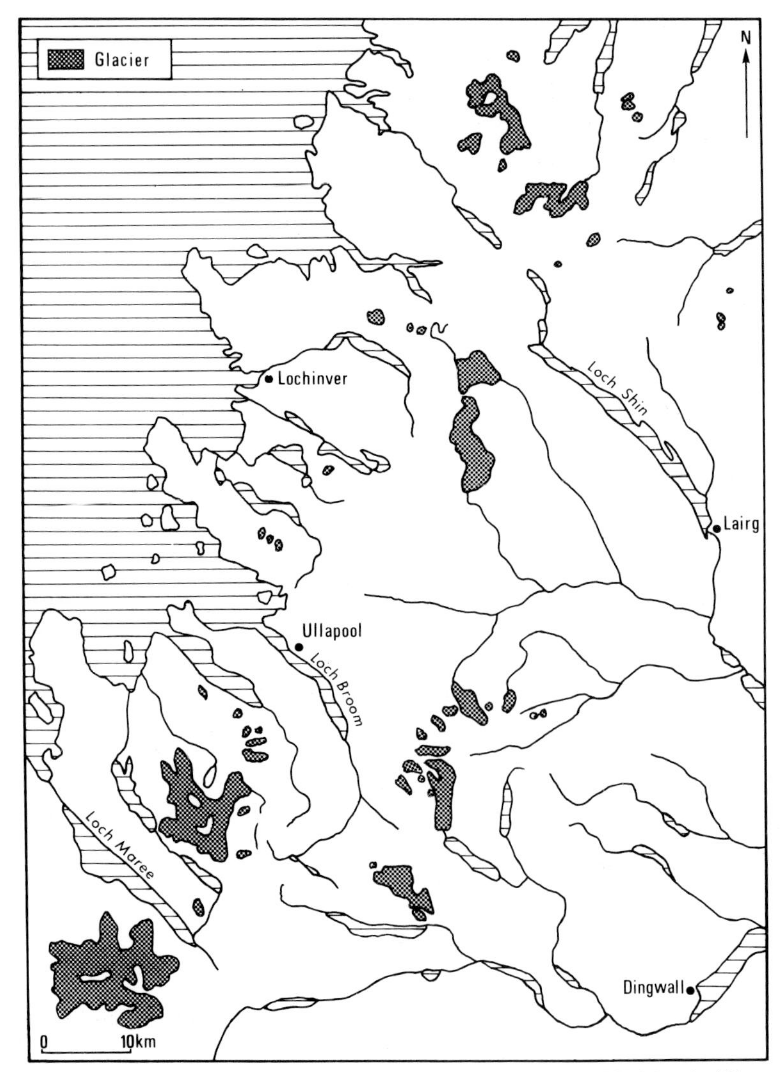

Fig. 4.26. Loch Lomond Advance glacier limits in the North-west Highlands (Sissons 1977b).

occupied Glen Geusachan with a lobe extending across the floor of Glen Dee down to an altitude of about 500 m. Another glacier (4 km long) occupied the valley on the east side of Cairn Toul and it descended to an altitude of 600 m. All the other glaciers which developed in the Cairngorms at this time were surprisingly small (less than 2 km long) and all of them occupied the heads of pre-existing corries. The distribution of the Cairngorm glaciers during the Lateglacial Stadial is surprising in three ways. Firstly, as this is a major area of high ground the limited size of the glaciers needs to be explained. Secondly, both the largest and the majority of the glaciers developed on the south and east sides of the mountain system. Thirdly, only small glaciers developed in the north facing corries. Sissons (1979b, p. 67), explains these characteristics as follows. "The sizes, distribution and altitudes of the glaciers and the absence of glaciers from some localities where they might have been expected to occur indicate that the principal snowfalls were associated with winds from southerly points (probably from south-east) and that south-west winds were important to transferring snow to some glaciers." It is inferred that snowfall diminished towards the north and north-west.

NORTH-WEST HIGHLANDS

The limit of the Loch Lomond Advance north of a line from Loch Torridon to the Cromarty Firth (Fig. 4.26) has been mapped by Sissons (1977b). Glaciers developed in mountain groups with considerable areas above 800 m. In the ground between Loch Torridon and Loch Maree a series of radiating glaciers up to 8 km long descended from an ice cap, the upper surface of which attained an altitude of 600 m. The outlet glaciers descended to sea-level in Loch Torridon and to altitudes between 100 and 400 m on the northern side. A similar glacier complex developed in the high ground between Loch Maree and Little Loch Broom but individual glaciers were only 5 km long and descended to 100 m on the north side. The four corries on the east face of Anteallach (1062 m) each contained individual glaciers one to three kilometres in length. Confluent valley glaciers developed on the north-east side of Sgurr Mor and on the east side of Beinn Dearg, in both cases descending to altitudes of 300 m. Ben More Assynt (998 m) nourished a 10 km long glacier which descended Glen Oykel to an altitude of 200 m and a broad lobe of ice descended the north face of this mountain. Small glaciers developed on individual mountains such as Quinag and Arkle although a series of confluent valley glaciers occupied the north-east face of Foinaven.

The distribution of Loch Lomond Advance glaciers in north-west Scotland as depicted by Sissons is rather surprising. He claims (Sissons 1977b, p. 57), ". . .it is believed that the great majority of the former glacier limits have been determined with considerable accuracy (1–100 m)." However, he does admit that, "Some former glaciers may not have been identified" and that ". . .it is possible that in some instances critical evidence is concealed by peat while in others glaciers may have failed to leave clear morphological evidence." There are certainly numerous valley systems in locations very close to the glacier systems recognised by Sissons which are similar in size, altitude and aspect and with what would appear to be adequate accumulation areas to support glaciers but which either do not contain evidence that such glaciers existed or contain evidence which has not yet been recognised.

Sissons explains the rather unusual distribution of ice masses which he has recognised in terms of a general decrease in precipitation inland from the west coast and superimposed upon this trend was a decrease in precipitation from south to north. The latter characteristic being suggested by some marked contrasts in glacier dimensions and altitudes on opposite sides of some of the individual mountain masses. The glaciers are frequently larger and reach low altitudes in south facing valleys.

THE WESTERN ISLANDS

The islands of Skye, Rhum, Mull and Arran all supported glaciers during the Loch Lomond Stadial (Figs. 4.27; 4.28). There were thirteen glaciers on Skye (Sissons 1977a), nine in the Cuillins and four in the eastern Red Hills. Five separate valley glaciers descended west and south-west facing slopes of the Cuillins and reached altitudes of 200 m. Glen Brittle contained a major glacier which terminated at about 100 m. The biggest glacier emanating from the Cuillins occupied Loch Coruisk and it was 7 km long and 2 km wide at the snout which probably calved into the sea. The four glaciers which developed in the Eastern Red Hills were all less than 2·5 km long and descended to altitudes of between 100 and 200 m.

The island of Rhum (Fig. 4.27), supported eleven glaciers, nine in the Rhum Cuillin and two in the western hills (Ballantyne & Wain Hobson, 1980). These glaciers were between 0·5 and 3 km long and descended to altitudes between 100 and 320 m although one glacier did descend to sea-level. The distribution patterns of glaciers on Rhum and Skye are very similar. Glaciers with a northerly aspect tended to be rather small whereas those with southerly or westerly aspects tended to be larger and descended to lower altitudes. East and north-east facing corries in both the Cuillins of Skye and the Rhum Cuillin nourished valley glaciers (in Glen Cuirisk and Glen Dibidil respectively) that flowed south-eastwards to below sea-level.

Mull developed a major ice cap (Fig. 4.28) with individual glaciers up to 12 km in length (Gray & Brooks 1972). Nunataks such as Beinn Na Duatharach (455 m), Beinn Talaidh (740 m) and Sgurr Dearg (740 m) stood above the ice surface which probably attained altitudes around 500 m. Outlet glaciers descended to sea-level in the Sound of Mull, Loch Spelve and Loch Buie.

A similar ice mass to that on Mull developed on Arran (Fig. 4.28) in the high ground of the north end of the island (Gemmell 1973). The summits of Goat Fell and Caistel Abhail stood above the ice surface but glaciers up to 6 km long radiated out from the high ground to terminate at altitudes of between 50 and 150 m above sea-level. It is possible that the Glen Rosa and Glen Sannox glaciers reached sea-level. The largest glaciers on Arran had a southerly or south-westerly aspect.

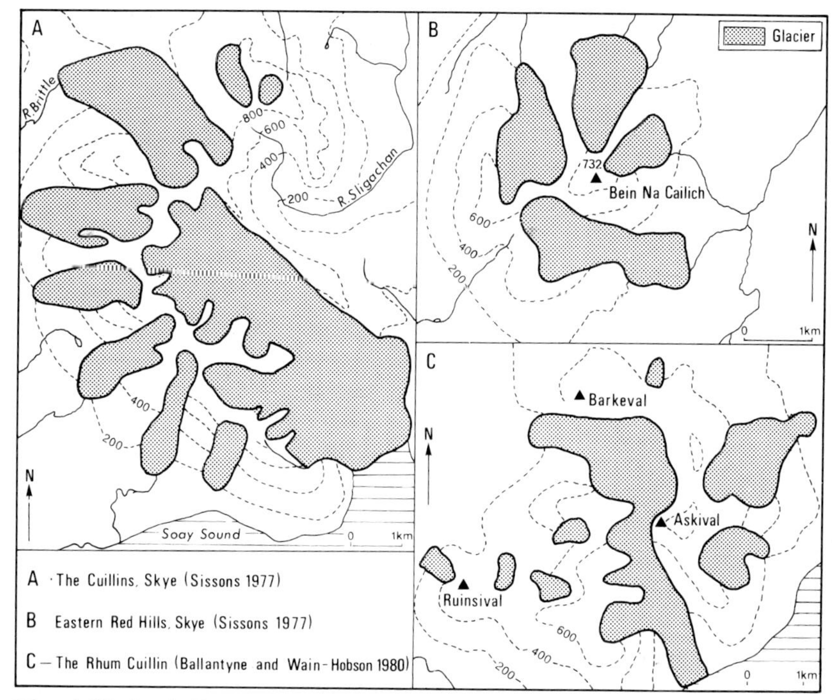

Fig. 4.27. Loch Lomond Advance glacier limits on the islands of Skye and Rhum. (Sissons 1977a, Ballantyne & Waine–Hobson 1980.)

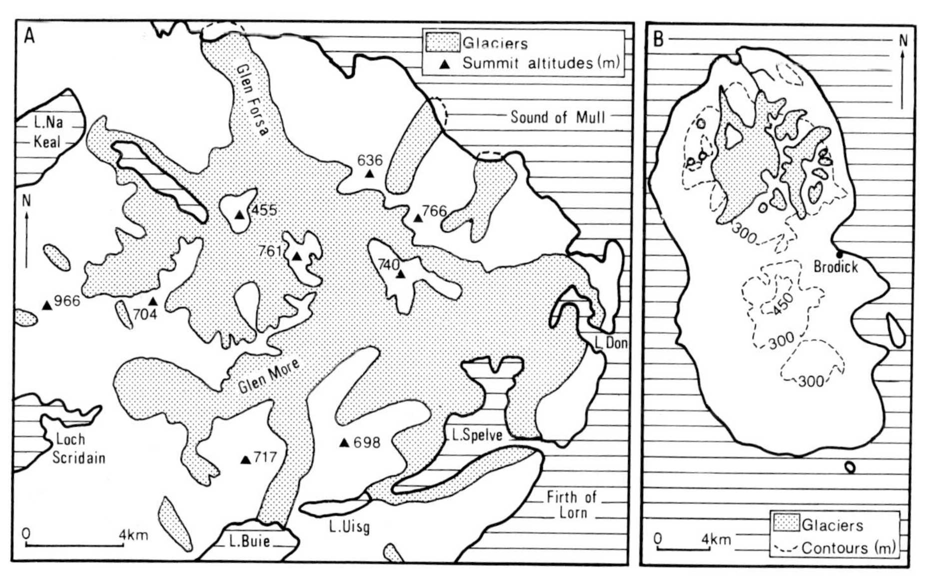

Fig. 4.28. Loch Lomond Advance glacier limits on the islands of Mull (Gray & Brooks 1972)—A. and Arran (Gemmell 1973)—B.

SOUTHERN UPLANDS

Glacier development during the Lateglacial Stadial was on a very limited scale in the Southern Uplands (Fig. 4.29). However, in 1864 Young mapped the extent of morainic mounds lying on top of the basal till in the Tweedsmuir Hills and this led J. Geikie (1894) to conclude that there had been a readvance of ice after the general retreat of the ice sheet. The extent of this hummocky moraine has been re-mapped

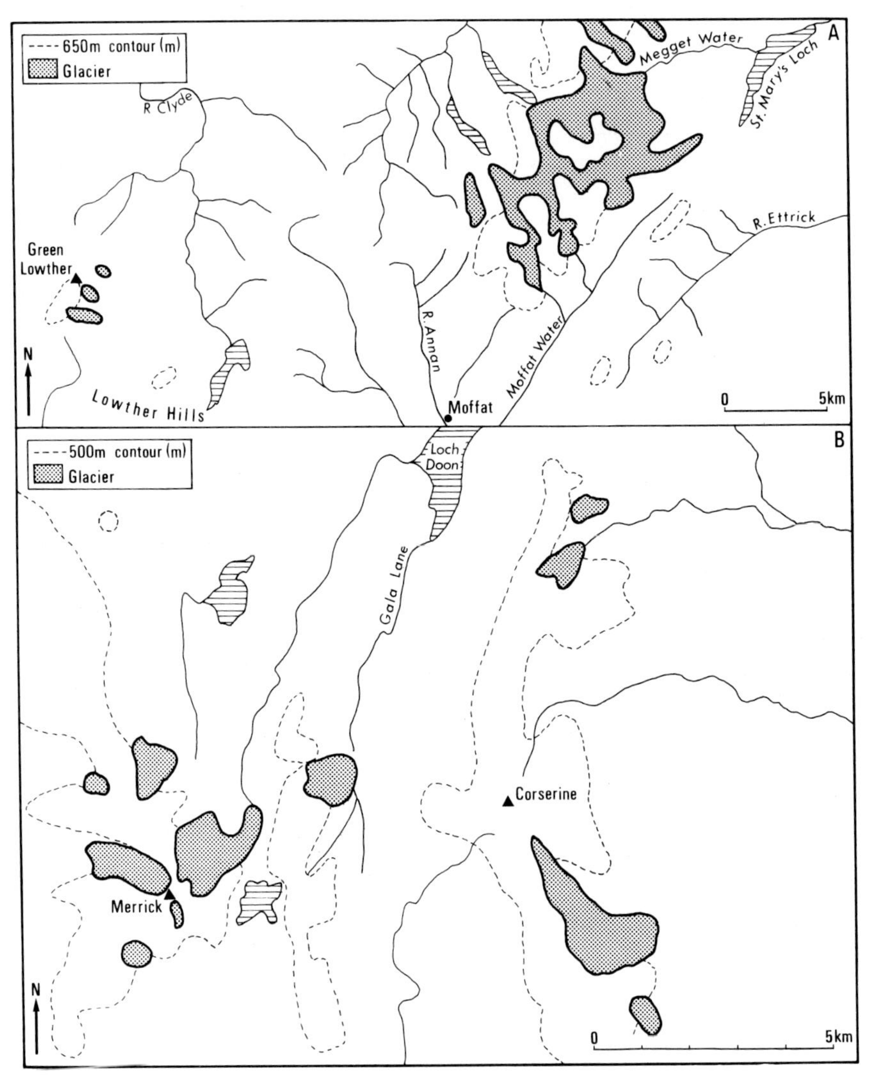

Fig. 4.29. Loch Lomond Advance glacier limits in (A) the Central Southern Uplands (Price 1963; Sissons 1967; May 1981) and in (B) the Western Southern Uplands (Cornish 1981).

by Price (1963) and May (1981) and it appears that a plateau ice cap (Fig. 4.29A) developed on the high ground between Broad Law and Hart Fell (generally above 700 m) from which a series of small valley glaciers descended to altitudes of between 300 and 400 m. The glaciers were longest (1·5 to 2 km) in valleys oriented towards the east and south-east.

Three small valley glaciers, less than 1 km long, developed in south-east facing valleys on Green Lowther (May 1981) and descended to an altitude of 400 m. A small ice mass also developed at the head of the Afton Valley (Holden 1977), with the main glacier descending the Afton Valley for 2 km and reaching an altitude of 330 m.

In the western Southern Uplands (Fig. 4.29B) the extent of eleven former glaciers has been established by Cornish (1981). Most of the glaciers occupied cirques facing between north and east and apart from two glaciers which covered areas of between 2 and 3 km^2 the remainder were less than a square kilometre in area. The mean value of the firn line altitudes which have been calculated for each glacier is 495 m.

Several general points emerge from the above consideration of the Loch Lomond Advance glaciers in Scotland. It must be stressed that the former extent of these glaciers is almost entirely based on landform assemblages with considerable emphasis on the distribution of 'fresh' hummocky moraine and moraine ridges, concentrations of small meltwater channels, fluted moraine and the absence of well-developed periglacial phenomena inside the postulated glacier margins. It is always assumed that the ice margins that have been deduced represent maximal extents of the glaciers during the Lateglacial Stadial and that these limits are contemporaneous. As more detailed investigations are undertaken it may well be found that some of these assumptions are incorrect.

It must also be remembered that the Loch Lomond glaciers mainly occupied cirques and troughs which were created during earlier periods of glaciation. It is likely that glacial erosion associated with the relatively short-lived (circa 1,000 years) Loch Lomond Advance glaciers was minimal compared to that achieved by the ice of the Devensian ice sheet and earlier glaciations.

Although our present knowledge of the extent of the Loch Lomond Advance glaciers indicates that a relatively small part of the total surface of Scotland was covered by active glaciers there can be little doubt that very large parts of the country above 400 m would have had a snow cover for large parts of the year and that large areas above 800 m probably had a permanent snow cover for several hundred years during the coldest part of the stadial.

PERIGLACIAL LANDFORMS AND DEPOSITS

In the previous section it has been demonstrated that extensive accumulations of glacier ice developed in Scotland during the Lateglacial Stadial and it is not surprising that the areas beyond and above these glaciers experienced intensive periglacial activity at the same time. Although periglacial processes are known to have continued to operate throughout the Flandrian, and are still active to-day, at high altitudes (Miller *et al* 1954; Ryder & McCann 1971; King 1972; Sissons 1979a) the lower limit of present day periglacial activity over mainland Scotland generally ranges between 500 and 600 m. Postglacial periglacial conditions appear to be relatively insignificant in Scotland compared with the intense periglacial activity which took place during the Loch Lomond Stadial and although many of these features are now fossilised they bear testimony to the severity of the climate at that time (Pl. 4.5A, B, C).

At low altitudes fossil frost wedges have been observed in deposits either laid down during the Lateglacial Interstadial or the early part of the Loch Lomond Stadial. Fossil frost wedges occur at 250 m just south of An Teallach in the north west Highlands, at two sites on Mull on low ground just inside the limit of the Loch Lomond Advance limit (Sissons 1976a) and in raised beach deposits on the north side of the Clyde estuary (Rose 1975). Péwé (1966) has suggested that frost wedge formation requires a mean annual temperature of at least −1°C and possibly lower than −6°C for their formation. Climatic conditions even near sea-level in Scotland were therefore very severe during the Loch Lomond Stadial.

These severe climatic conditions are recorded at numerous sites in which Lateglacial pollen investigations have been undertaken. The relatively low pollen counts obtained for the period equated with the Loch Lomond Stadial are always associated with minerogenic materials reflecting the periglacial processes active on the surrounding slopes (Lowe & Walker 1977; Pennington 1977; Vasari 1977). A layer of solifluction gravel at Bigholm Burn, Dumfriesshire at an altitude of 160 m has been dated to between 11,500 and 9,500 bp (Bishop & Coope 1977). At two sites in the Glasgow area soliflucted till moved down the lower slopes of drumlins sometime after 11,500 bp (Dickson, Jardine & Price 1976).

It is within the Highlands, however, that there is a great deal of evidence relating to intense periglacial activity during the Loch Lomond Stadial. Sissons (1974c) has described major solifluction lobes on the Lochnagar and Mount Keen granite areas and large solifluction sheets in the Gaick area with fronts from 2 to 7 m high. He points out that it is significant that the solifluction lobes and sheets occur down to altitudes of about 600 m but that large areas at much higher altitudes (up to 900 m) lack such features. Throughout the Highlands Sissons has demonstrated that well-developed periglacial deposits and forms are absent from areas occupied by the Loch Lomond Advance glaciers.

In the Cairngorms, as well as solifluction lobes up to 10 m high (Sissons 1979b), fossil rock glaciers with curving terminal ridges resembling end moraines have been described. Similar features have been described in Jura (Dawson 1977) and Wester Ross (Sissons 1975b). Deposits associated with perennial snow beds are much more common. Protalus ramparts occur in the Cairngorms and the north-west Highlands. A very large protalus rampart has been described in Wester Ross (Sissons 1976b) which is 1 km long and up to 55 m high, its volume implying a 17 m retreat of the rock face that supplied it.

In geomorphological terms the Lateglacial Stadial was a very short interlude but the intensity of periglacial activity, even at low altitudes, was certainly great. To what extent the landforms produced by the Devensian ice sheet were modified by these periglacial processes is difficult to quantify (Price 1980). However, there can be little doubt that significant slope modification took place and that weathering processes made ready considerable amounts of material which were to be subsequently transported down slope during the early part of the Flandrian.

Plate 4.5. Blockfield (**A**), Stone Stripes (**B**) and Stone Circles (**C**) on the western hills of Rhum (C. Ballantyne).

COASTAL LANDFORMS AND DEPOSITS

The evidence that has been studied relating to relative sea-level during the Loch Lomond Stadial is limited to three areas: the Firth of Forth (Sissons 1974d, 1976c), the Firth of Clyde (Peacock *et al* 1977; Peacock *et al* 1978; Gray 1978) and the shores of the Firth of Lorn and the Sound of Jura (Gray 1974a, b, 1975a, b, 1978; Dawson 1980a, c). The sea around Scotland at this time reflected the cold environment of the Stadial in that deposits accumulated in the inshore zone contain foraminifers, ostracods and dynoflagellate cysts which indicate cold or very cold conditions (Binns *et al* 1974; Peacock 1974, 1975; Peacock *et al* 1978). These cold water deposits overlie the interstadial Clyde Beds, the distribution of which has been much modified by erosion along the coastlines of central Scotland during the Loch Lomond Stadial.

Continued isostatic uplift during both the Lateglacial Interstadial and Lateglacial Stadial resulted in continuous fall of relative sea-level. Sissons (1969, 1974d) describes an area near Grangemouth at least 28 km^2 in extent which was planated

Plate 4.6. Raised marine platform and cliff—Lismore Island (Inst. of Geological Sciences photograph published by permission of the Director, N.E.R.C. copyright).

by the sea across till and bedrock, and Lateglacial marine deposits, close to Ordnance Datum. Farther east at Leith the platform is cut in till and the shoreline is at −5 m. This shoreline continues to decline in altitude eastwards and reaches −18 m, 9 km north of Berwick. It is therefore likely that much of the east coast of Scotland was experiencing a period of low relative sea-level during the Loch Lomond Stadial with the shoreline being at altitudes of between −5 and −20 m O.D.

Detailed work by Gray, Peacock, and Sutherland along the coastline between the Firth of Clyde and Mull has revealed that along much of this coast, sea-level stood at from 5 to 10 m above present Ordnance Datum during the Loch Lomond Stadial. On the eastern coast of Mull, in the Oban area and on the shores of the Sound of Jura there are some very well-developed raised shore platforms and cliff lines (Pl. 4.6). Many workers believed these features to be Holocene in age because they correspond in altitude with Holocene raised beaches. Wright (1928), for example, called the Holocene sea "The cliff maker par excellence." McCallien (1937), however, suggested that the Holocene epoch was too short to allow the necessary erosion, and instead proposed a 'pre' or 'inter' glacial age, a suggestion that subsequently became generally accepted (McCann 1966a, 1968; Synge 1966; Synge & Stephens 1966; Sissons 1967a; Gray 1974a).

In the Firth of Lorn area it has been shown (Gray 1974a) that there is one major erosional shoreline present which is tilted in a westerly direction from 11 m O.D. north of Oban to 4 m O.D. on mid-Mull and slopes southwest to descend below present sea-level on the Mull of Kintyre (Fig. 4.30). Sissons (1974a, 1976c) suggested that this shoreline can be correlated with the Lateglacial erosional feature in the Firth of Forth which declines in altitude eastwards from about O.D. at Grangemouth to −18 m O.D. near Berwick.

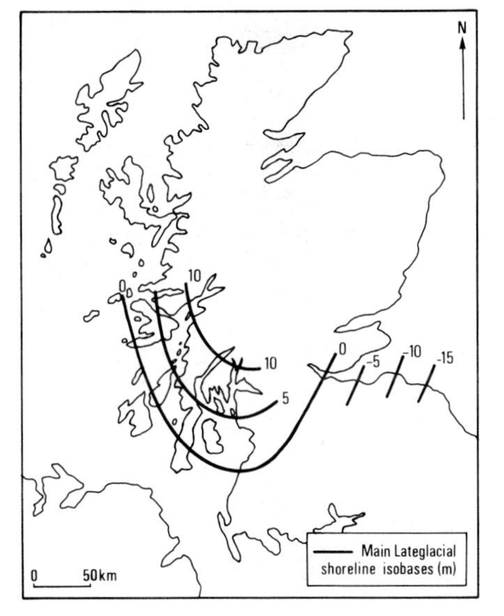

Fig. 4.30.

Isobases (m) of the Main Lateglacial Shorelines (Gray & Lowe 1977).

A major problem remains in explaining the formation of these erosional platforms in a relatively short period of time (circa 1,000 years) and at a time when it is believed that both isostatic and eustatic changes were taking place. In order for the platforms to develop the rate of rise of the land due to isostatic recovery must have equalled the rise in sea-level in those areas where the platforms developed. Such a situation has been demonstrated (Fig. 4.31) at Lochgilphead and Ardyne Point (Peacock *et al* 1977, 1978) where sea-level was relatively stable at 7 to 10 m O.D. from 11,500 bp to 10,300 bp.

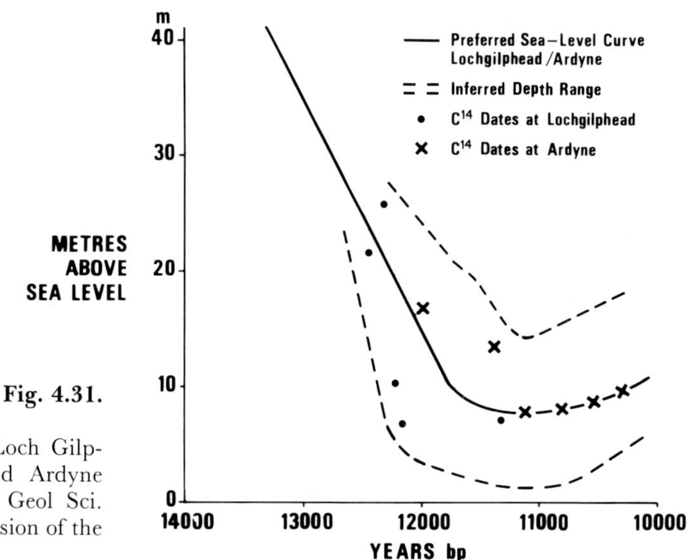

Fig. 4.31.

Lateglacial sea-level curve for Loch Gilphead (Peacock *et al.* 1977) and Ardyne (Peacock *et al.* 1978). (Inst. of Geol Sci. diagram published by the permission of the Director, NERC copyright).

Even with a relatively stable sea-level it is hard to envisage in what manner many of the rock platforms could have been cut within a period of about 1,000 years. Many of them are cut in solid rock in situations with very limited fetch for wave development. Sissons (1974d) suggested that such rapid platform cutting was related to the intense periglacial environment of the Loch Lomond Stadial and this suggestion was taken up by Gray (1978, p. 161) and he believes that the combination of periglacial weathering, ice foot activity, marine erosion and removal of debris in what is believed to have been a period of storm conditions (Sissons & Sutherland 1976) led not only to the rapid formation of the platforms but to their formation in situations in which marine erosion alone would not have produced them.

Although accurate information about relative sea-level is limited, the tentative construction of isobases for the Main Lateglacial Shoreline (Fig. 4.30) would suggest that the Scottish shoreline, apart from that segment between Ayr and Loch Torridon, was generally 5 to 15 m lower than at present. The correlation of relatively higher shoreline altitudes at the time of the Loch Lomond Stadial with those areas most affected by glacier development cannot be ignored. Perhaps the rate of isostatic rebound of the west central Highlands was slowed down by the build up of glacier ice in that area.

VEGETATION AND SOILS

In numerous pollen diagrams illustrating spectra from the period of the Loch Lomond Stadial in Scotland the climatic deterioration is represented by lower pollen concentrations, poorer pollen and spore preservation and a return to predominantly coarse, minerogenic sedimentation. At many sites aquatic pollen is absent or at least there are marked changes in the records of macro fossils of aquatic plants. The vegetation of Scotland at the time was open tundra with only occasional tree birch although dwarf birch was more common (Vasari & Vasari 1968; Pennington *et al* 1972; Birks 1970, 1973; Walker 1975; Bishop & Coope 1977; Lowe & Walker 1977; Pennington 1977a, b; Walker & Lowe 1977; Birks & Matthewes 1978; Walker & Lowe 1979; Vasari 1977; Caseldine 1980; Macpherson 1980). Over the whole of Scotland there was a dramatic break up of the vegetation cover and destruction of the soils that had developed during the preceding Interstadial. The generally low carbon content of Stadial sediments reflects the destruction of the Interstadial soils.

Dominance of pollen of *Compositae, Cruciferae, Caryphyllaceae* and *Chenopodiaceae* reflect open ground and disturbed soils. *Rumex* was abundant and *Artemisia* was especially characteristic (Fig. 4.32). The fact that pollen of plants which favour base-rich soils, are common, suggests large areas of fresh mineral soils subject to freeze-thaw processes.

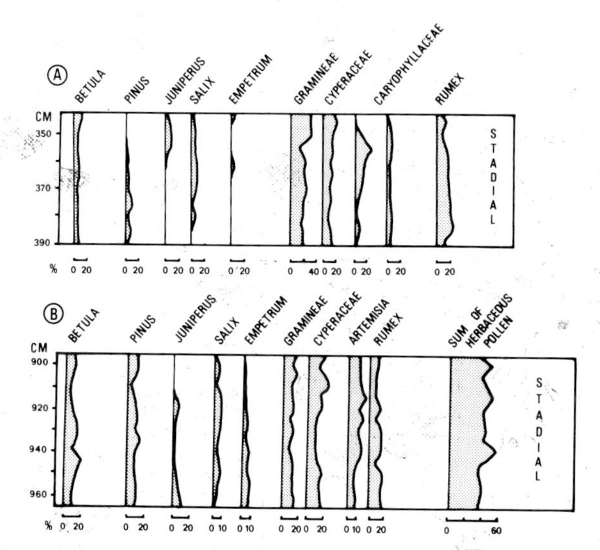

Fig. 4.32. Percentage pollen diagrams (selected taxa) from (A) Tirinie and (B) Amulree, Eastern Grampians (Lowe and Walker 1977).

The large amounts of *Artemisia* present in many pollen profiles presents certain problems. It is a genus which is usually associated with continuously frost-heaved soils of a continental periglacial environment (Iversen 1954). Macpherson (1980, p. 98) has mapped the geographical variation of Loch Lomond Stadial pollen spectra (Fig. 4.33) and points out that "North and east of the main watershed . . . values of *Empetrum* plus *Salix*, together with those for other woody taxa and

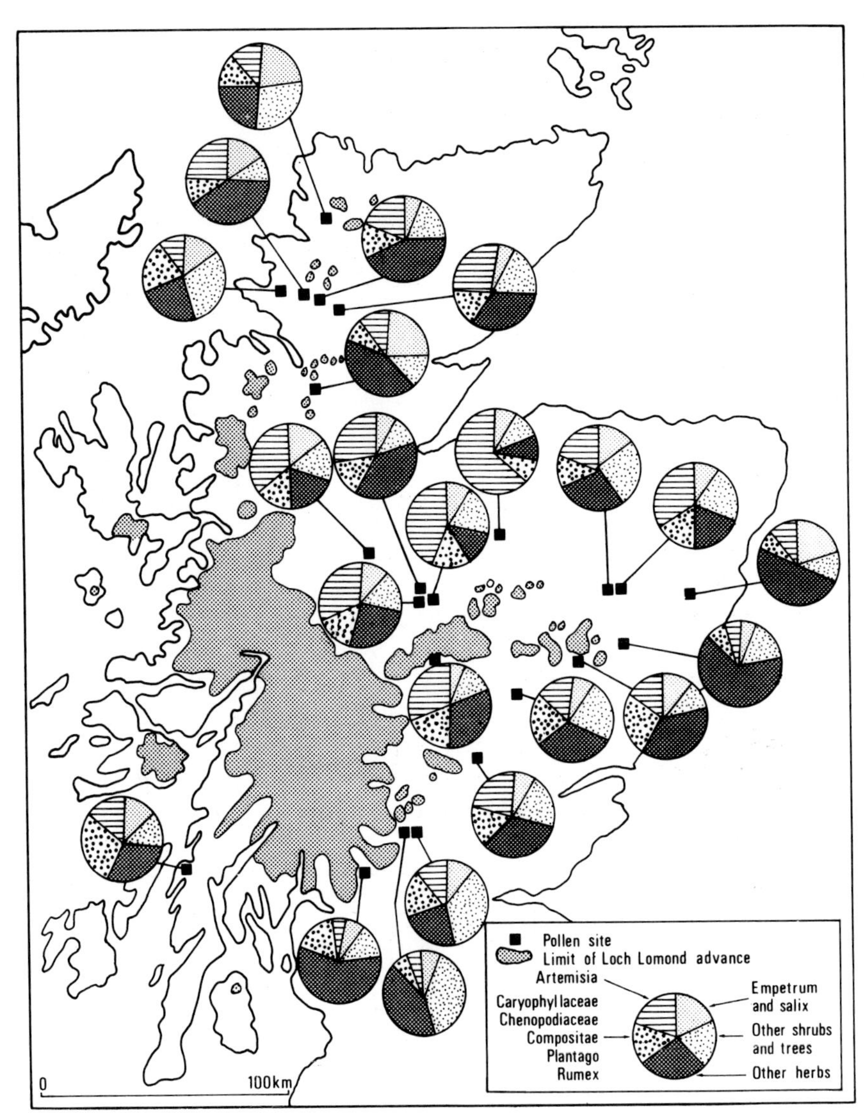

Fig. 4.33. Pollen spectra at *Artemisia* maximum from sites in and adjacent to the Scottish Highlands (MacPherson 1980).

'other herbs' are generally lower than at the south and north-west margins of the Highlands." *Artemisia* values, by contrast, are generally higher in the lee of the watershed, reaching over 60 per cent at the Abernethy Forest site. These high values led Birks & Matthewes (1978) to suggest low precipitation in this area of the Highlands; Sissons' suggestion (1980) of an annual precipitation of only 200–300 mm in Speyside lends support to their views.

It does appear therefore that there were important regional variations in climate during the Loch Lomond Stadial in Scotland which are reflected in plant distributions. However, the landscape was dominated by grasses, sedges and shrubs and soil development was minimal. As the Stadial came to an end *Empetrum* and *Juniperus* heath expanded quickly and birch woodland was widespread in Scotland within about 1,000 years.

FAUNA

As might be expected, the terrestrial faunal evidence relating to the Loch Lomond Stadial is rather limited. Insect assemblages from Redkirk Point (Bishop & Coope 1977) indicate the deterioration of climatic conditions at the beginning of the Stadial with a consistent increase in the numbers of the northern species and in those of more open habitats but there is an absence of characteristic species of the alpine zones or of the true tundra, and Bishop and Coope suggest that the average July temperature about 11,000 bp was 10°C. But evidence from another site (Bigholm Burn B) in the form of the presence of the beetle *Olophrum boreale* and the crustacean *Lepidurus arcticus* suggest much more severe conditions.

The vertebrate population of Scotland at this time included reindeer, arctic lemming, Norway lemming, tundra vole and ptarmigan (Stuart 1977). The remains of these animals certainly bear testimony to the sparse nature of the vegetation and to the low temperature levels of the Stadial.

The inshore waters around Scotland at the time of the Loch Lomond Stadial were certainly much colder than at present. A borehole east of Colonsay (Binns *et al* 1974), a borehole in the Cromarty Firth (Peacock 1974) and excavations at Ardyne Point (Peacock *et al* 1978) all reveal low- or mid-arctic conditions during the period 11,000 to 10,000 bp. At Ardyne Point the occurrence of such species as *Portlandia arctica* and *Arctica islandica* indicates a low Arctic environment while the dominance in the foramaniferal species of *Elphidium barletti*, *Elphidium clavatum* and *Elphidium subarcticum* suggests Arctic and Subarctic conditions. The ostracod population is dominated by Subarctic to Boreal species. The climatic deterioration suggested by the build up of glacier ice on the land and the development of tundra-like soils and vegetation was certainly paralleled by cold waters around the Scottish coasts.

CLIMATE

The deterioration of climatic conditions during the Loch Lomond Stadial is clearly indicated by the geomorphological, lithological, botanical and faunal evidence discussed above. Even though the radiocarbon dating of the limits of the Stade may be open to question there can be little doubt that it was a relatively short period of deterioration. The main outstanding question is the extent to which temperatures declined, because the development of glaciers, frost wedges, a tundra-like vegetation and 'cold' terrestrial and marine faunas all indicate a climate considerably colder than the present.

The first temperature curve covering the Loch Lomond Stadial time period in Scotland was published in 1977 (Bishop & Coope 1977, Fig. 11) and was inferred from coleopteran assemblages from sites in south-west Scotland (Fig. 4.11). The average July temperature begins to fall about 11,000 bp from 12°C and continues to decline to below 9°C at about 10,200 bp. It then begins to rise steeply and reaches 15°C by 9,700 bp. At one of the sites at which the beetle evidence was studied, Bigholm Burn, there was evidence of solifluction and therefore there must have been quite a large annual temperature range if the July average was about 9°C. There can be little doubt that much of Scotland, outside the areas covered by the Loch Lomond Advance glaciers, experienced discontinuous permafrost at some time during the Stadial. Since such conditions imply a mean annual temperature no higher than −1°C it is likely that mean January temperatures were of the order of −7°C at sea-level. An annual temperature range of some 16°C is some 50 per cent greater than the present range and implies very severe temperature conditions on high ground during the winter. Assuming a decrease in temperature of 0·6°C/100 m, a mean January temperature no higher than −17°C is indicated on the top of Ben Nevis.

All the evidence therefore points to very severe winters over a period of several hundred years and this, of course, explains the build up of glaciers in the western Highlands and other high ground in Scotland. Sissons & Sutherland (1976) have attempted an ambitious reconstruction of Scotland's climate during the Loch Lomond Stadial on the basis of the reconstruction of ice marginal positions allowing the thickness of individual glaciers to be estimated and their mass balance to be calculated at their maximum extent. Such calculations require various assumptions to be made in terms of precipitation, temperature and ablation rates. The glacier limits are themselves a subject of controversy and some of the assumptions in the mass balance equations may also be questioned. Nevertheless, the conclusions reached by Sissons and Sutherland (1976) and Sissons (1980) are extremely interesting and so long as the dangers of circular arguments are borne in mind the following summary of those conclusions remains the most exciting reconstruction of Scotland's climate at any period before climatological records are available.

The distribution of the Loch Lomond Advance glaciers (Fig. 4.12) reveals a strong concentration of ice in the western Highlands and a surprisingly small amount of ice in the eastern Grampians, Cairngorms, Monadhliath Mountains and North-west Highlands. Such a distribution might suggest that the principal snow-bearing winds were from the west or south-west. It might also be expected that glaciers occupying north and north-east sites would have been larger and reach lower altitudes due to the effect of direct insolation. However, Sissons (1980) has demonstrated that both these assumptions are invalid. His maps of the individual ice masses frequently demonstrate that glaciers oriented between south-east and west are larger, longer and reach lower altitudes.

Sissons (1980, p. 35) has produced a map of regional firn line altitudes based on 226 individual values in the Highlands and Inner Hebrides (Fig. 4.34). There are two major trends revealed by this map. Firstly, the regional firn line rose from 300/400 m in the south-west and west to over 800 m in the Cairngorms and Monadhliath Mountains: that is, there was a marked rise from the west coast, inland. Secondly, there was a marked rise in the regional firn line from south-south-east to north-north-west from the south-east edge of the Grampian Highlands. Sissons explains these two trends in terms of the importance of both south-westerly

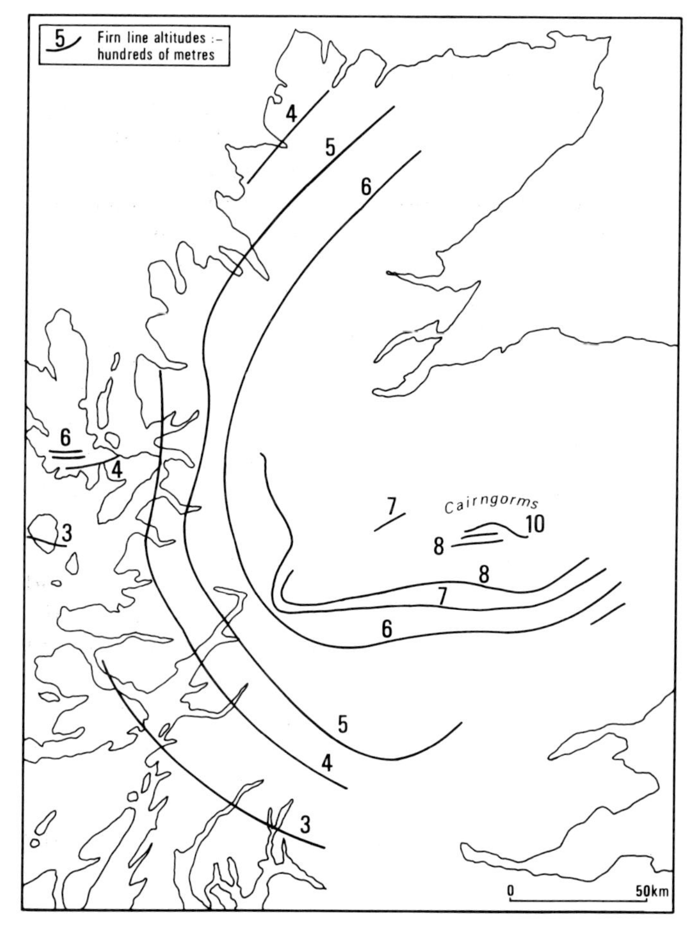

Fig. 4.34.

Regional firn line altitudes (Sissons 1980b).

and south-easterly air streams and the effect of 'snow'-shadow in the form of the expanding ice mass in the west on the one hand and the barrier of the Highland 'edge' to south-easterly winds on the other. It is therefore possible to regard the lines on Figure 4.34 as very general isohyets with values inversely related to the firn line altitudes.

This precipitation pattern is explained by Sissons and Sutherland (1976) in terms of the movement of depressions along more southerly routes than they follow to-day. They stated (1976, p. 345), "It appears that the only way in which the climate of the Stadial in Scotland can be explained is by invoking the return of the cold polar water" and "it may well be that the junction of polar and relatively warm (North Atlantic Drift) waters was in the immediate vicinity of the British Isles during the Stadial. In such a situation the contrasting sea temperatures would have encouraged the vigorous interaction of air masses that would have resulted in stormy conditions and abundant precipitation in much of the British Isles." Ruddiman *et al* (1977) state that the boundary of polar and relatively warm water was located at the latitude of south-west Ireland at this time.

Sissons estimated precipitation values by calculating the average net accumulation on a glacier and by determining the area above the firn line and the glacier's volume and assuming a time period of 750 years (Sissons 1980b, p. 40). Such calculations suggest that the south-west Grampians received four to six times as much precipitation as the Cairngorms. Taking other factors into consideration Sissons concludes that reasonable precipitation estimates for the two areas would be 3,000–4,000 mm and 500–600 mm respectively at about 1,000 m above sea-level. He suggests a figure of 200–300 mm on the low ground of Speyside. These low precipitation figures for the north-eastern area are explained by the more southerly track of depressions, the low air temperature and by the elimination of the North Sea as a major source of moisture as large parts of it would have almost certainly been frozen over in winter. It would seem, therefore, that Speyside and the north-east of Scotland experienced drier (less snow fall) and much colder conditions than the western Highlands. This conclusion is borne out by the distribution of *Artemisia* as revealed by pollen analyses (Pennington *et al* 1972; Walker 1975a, 1975b, 1977; Lowe 1978; Lowe & Walker 1977; Birks & Matthewes 1978; Macpherson 1980). It has been suggested that species of *Artemisia* prefer dry ground and limited snow cover. Expressed as a percentage of total land pollen at sites in the north-west Highlands and along the south-east Highland edge *Artemisia* represent 10–15 per cent of the total pollen but at sites in the Great Glen the value rises to between 30 and 65 per cent. A study of the distribution of *Empetrum* and *Salix* at the horizon of the *Artemisia* maximum (MacPherson 1980) also supports the conclusion that the north-eastern Highlands were drier than the western and southern Highlands. MacPherson (1980, p. 98) states "Thus the pollen data reveal a coherent geographical pattern the climatic implications of which are consistent with the evidence provided by firn-line calculations."

The probable climatic conditions during the Loch Lomond Stadial (Table 4.3) based on the best available evidence can be estimated as follows:

TABLE 4.3

Estimated temperature and precipitation values during the Loch Lomond stadial in Scotland

		Av. January Temp °C	Av. July Temp °C	Annual Precipitation mm.
South West				
Scotland	sea-level	−7°	9°	1,500
Western	sea-level	−9°	7°	1,500
Grampians	1,000 m	−17°	0°	3,000
North East	sea-level	−12°	6°	200–300
Highlands	100 m	−18°	0°	500–600

The regional and altitudinal variations in climatic conditions during the Loch Lomond Stadial were certainly sufficient to produce variations in the vegetation conditions. The environment was also characterised by rapid changes in all its aspects and it is probable that at no other time during the last 30,000 years was the rate of change in environmental conditions so great. It is really quite remarkable that temperature and therefore other aspects of the environment should change so radically within a period of approximately 1,000 years.

CONCLUSIONS

The Lateglacial period in Scotland has been shown to be characterised by great changes in the environment. After the rapid wastage of the Devensian ice sheet the land was colonised by plants and animals in what was a climate not very different from that of to-day. Over a period of about 1,000 years an open vegetation of grass, heath and shrubs developed, but there was insufficient time for trees to colonise Scotland before the temperatures began to drop. Within another thousand years temperatures dropped even further and permanent snow beds and glaciers developed. Throughout the Lateglacial period relative sea level was generally falling as a result of the isostatic rebound of the land taking place faster than the eustatic rise in sea level. This situation was halted in some areas during the Loch Lomond Stadial when sea level was stable. The thousand years of Stadial conditions were both dramatic in their impact upon the land, vegetation and fauna and they represent a temporary but final return to glacial conditions. All the evidence points to a very abrupt termination to this glacial interlude and it is probable that the change to climatic conditions not very dissimilar to those of to-day was achieved in about 500 years.

It is not the purpose of this book to consider all the possible causes responsible for changes in environmental conditions. However, the dramatic changes which took place in Scotland during the Lateglacial period can be linked with major changes in oceanic conditions to the west of the British Isles. Ruddiman &

McIntyre (1981) have determined the location of the North Atlantic polar front which separates polar (cold) from subpolar (relatively warm) waters at various dates (Fig. 1.10). About 13,000 bp the polar front was located off north-west Spain but it rapidly retreated north-westwards so that between 13,000 and 11,000 bp it was located to the south-west of Iceland. The return of the North Atlantic Drift to the shores of Britain and Scandinavia is closely associated with the wastage of the Devensian ice sheet and the establishment of interstadial conditions. The polar front then pushed south-eastwards again and lay to the south of Ireland between 11,000 and 10,000 bp before retreating north-west again to south-east Greenland by 9,000 bp. This brief return of polar water to the shores of Scotland coincides with the Loch Lomond Stadial and the Loch Lomond Advance glaciers.

The sequence of environmental changes which has been recognised in Scotland during the Lateglacial period and which may be summarised as a period of climatic amelioration followed by a brief revertence to 'glacial' conditions has been described by Watts (1980) as "a false start to the present Interglacial". Similar 'false starts' have been recognised in England, Ireland, the Scandinavian countries, and north-west continental Europe. The Lateglacial cooling is much less significant in central Europe and is recorded in oceanic sediment cores from the eastern but not the western North Atlantic ocean. Evidence for Lateglacial climatic oscillations in other parts of the world which can be correlated with those in Europe is not very satisfactory (Wright 1937). Watts (1980, p. 20) also points out that "The Flandrian appears to be the only interglacial cycle immediately preceded by a complex climatic event like the Lateglacial."

5

THE EARLY POSTGLACIAL PERIOD circa 10,000–5,000 bp

The rapid termination of the Lateglacial cold period resulted in the establishment of a temperate environment in Scotland very similar to that of the present. Not since the Ipswichian Interglacial some 120,000 years ago had Scotland enjoyed an extended period (i.e. more than 3,000 years) of temperature conditions conducive to the recolonisation of the land by forests. The rapid environmental changes of the Lateglacial period were replaced by a period of relative stability in all aspects of the environment. It can in fact be argued that during the Postglacial period the activities of man have produced greater changes in the physical environment (see Chapter 6) than those achieved by natural processes. There seems little doubt that for the first time in the last 30,000 years periglacial and/or glacial processes ceased to be major modifiers of the landscape and the processes of erosion on the land were probably slowed down by the development of an extensive vegetation cover. The continued wastage of ice sheets in other parts of the world and the remnant isostatic rebound of parts of Scotland did produce changes in the position of the Scottish coastline during the Postglacial period. However, in great contrast to the Lateglacial period temperature remained remarkably stable, only fluctuating by one or two degrees over the last 10,000 years. This relatively stable climate allowed the rapid spread of a forest cover which, if it had not been for the subsequent arrival of a human population which removed many of the trees, would still form a major element in Scotland's environment.

LANDFORMS AND SEA-LEVELS

There is surprisingly little known about the changes which have taken place in the landforms of Scotland during the last 10,000 years. The processes of slope modification and fluvial erosion, transportation and deposition in an essentially temperate environment must have modified considerably both the erosional and depositional landforms produced by the glacial and periglacial processes of the previous thirty thousand years. The only effective way of gauging the efficiency of the fluvial system during the Postglacial period is to examine the sedimentary record preserved in lake basins. Such work is in its infancy in Scotland (Pennington 1972; Dickson *et al* 1978; Edwards and Rowntree 1980) and little can be said with confidence about rates of erosion during the period 10,000 to 5,000 bp. It has been suggested that the presence of a closed forest cover over much of Scotland during this period (see next section) resulted in increased evaporation rates which more than compensated for the increased precipitation which is believed to have occurred between 7,500 and 5,000 bp. If this assumption is correct, then lower run-off conditions would have prevailed and this might explain the relatively low rates of sediment yield which have been suggested for this period (Edwards & Rowntree 1980).

It will be demonstrated later in this chapter that from about 8,500 bp relative sea-level was rising along most of the Scottish coastline and such a situation must have had an effect, at least in the lower reaches of the river systems. The period of dissection of valley floors and the production of terraces which characterised the Lateglacial period in Scotland must have been replaced by some aggradation or at least the slowing down of the excavation of the drift-filled valley floors.

The most significant landform changes to occur between 10,000 and 5,000 bp in Scotland took place along the coastline. During the last twenty years a reappraisal of the so-called '25 foot' raised beach (also referred to as the Neolithic beach) has taken place. Work by McCann (1961a, b, 1964, 1966) and Synge & Stephens (1966) on the west coast, by Sissons (1962, 1966) and Cullingford & Smith (1966) on the east coast and by Jardine (1962, 1963, 1964) in Ayrshire and the Solway Firth laid the foundations for a series of research projects which approached the problems of Postglacial sea-level changes in a rigorous manner. The results of these investigations will be discussed below in their regional contexts but it will be useful to summarise at the outset the general conclusions that have been reached about fluctuations in relative sea-level between 10,000 and 5,000 bp.

Sissons (1976) has produced a map showing the isobases of the Main Postglacial Shoreline (Fig. 5.1). Although this map will no doubt be modified as more data become available, it does reveal that the shoreline produced by the major marine transgression (the Holocene/Flandrian transgression) has, since its formation, been uplifted and warped around a centre of uplift in the general vicinity of Rannoch Moor. The isobases have an ellipsoidal form. The highest known altitude of the shoreline, 15 m, has been recorded at the western end of the Forth Valley but the shoreline reaches 14 m in the Firth of Clyde, in the Loch Lomond basin, on the shores of Loch Etive and Loch Linnhe. Although the whole of mainland Scotland was affected by this marine transgression the geomorphological effects of the transgression have not yet been deciphered along the north-east and north coasts and are only poorly understood in the Outer Hebrides (Ritchie 1966) and in Orkney and Shetland (Hoppe 1965).

The Main Postglacial Shoreline has two principal expressions in the Scottish landscape. It is either represented by raised sand and gravel beaches (sometimes resting on a rock-cut platform of Lateglacial or Interglacial age) or by raised estuarine mud flats known in Scotland as 'carselands'. The raised sand and gravel beaches occur on the more exposed parts of the coastline while the carselands occur in the more sheltered firths of the Forth, Tay, Upper Clyde and Solway. It has been very fortuitous that the relatively low relative sea-levels which occurred in these firths at the beginning of the Postglacial period allowed peat to accumulate on surfaces which were subsequently invaded by the sea as the marine transgression produced higher relative sea-levels. The burial of these peats by the carse clays has allowed the dating of the beginning of the transgression and the subsequent growth of peat on top of the carse clays as the sea retreated has allowed the dating of the marine regression. Such a chronology has now been established for the Solway Firth (Jardine 1975), the western Forth Valley (Sissons & Brooks 1971) and Lower Strathearn, Tayside (Cullingford *et al* 1980). The sea-level curves for these three areas (Fig. 5.2) show that relative sea-level was low between 9,000 and 8,000 bp but began to rise rapidly to reach a maximum during the Postglacial period about 6,600 bp in the Forth Valley, about 6,100 bp in the Tay

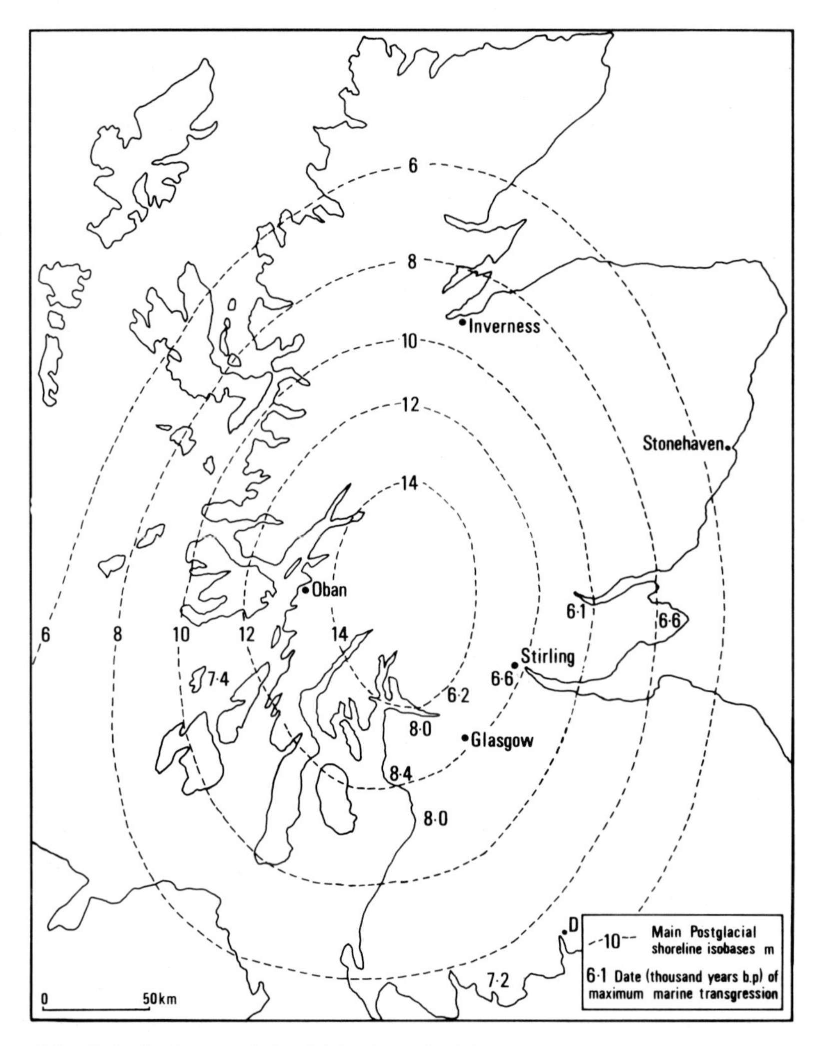

Fig. 5.1. Isobases of the Main Postglacial shoreline (Jardine 1982) and dates of the maximum Postglacial marine transgression (various sources).

Valley and at about 5,000 bp in Wigtown Bay on the Solway Firth. These curves show that relative sea-level rose by some 8 m between 8,500 and 7,000 bp in the Forth Valley and by a similar amount between 7,800 and 6,500 bp in the Tay Valley. Along the north shore of the Solway Firth relative sea-level began to rise

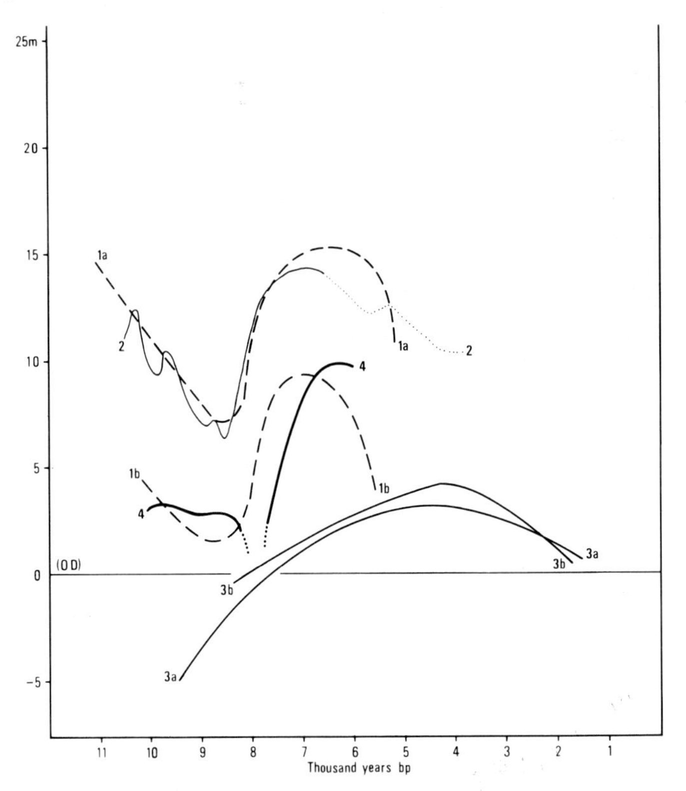

Fig. 5.2. Sea-level curves for: 1a, b—Central Scotland and 'marginal areas of Scotland' (Donner 1970); 2—Western Forth Valley (Sissons & Brooks 1971); 3a, b—Solway Firth (Jardine 1975b); 4—Lower Strathearn (Cullingford *et al.* 1980). All re-drawn to same scale (Jardine 1982).

somewhat earlier with a total rise of 8 m occurring between 9,200 and 6,000 bp. Whereas relative sea-level began to fall after 6,600 bp in the Forth Valley, and after 6,100 bp in the Tay Valley the regression did not begin until after 5,000 bp on the north shore of the Solway Firth.

The main Postglacial marine transgression in Scotland can be explained by the rate of the general rise in world sea-level, caused by the melting of the great North American and Scandinavian ice sheets between 10,000 and 5,000 bp, being greater than the continued rise of the land in response to the unloading by glacier ice prior to 10,000 bp. By using published curves of the eustatic rise in sea-level, both Sissons & Brooks (1971) and Cullingford *et al* (1980) have calculated land uplift curves

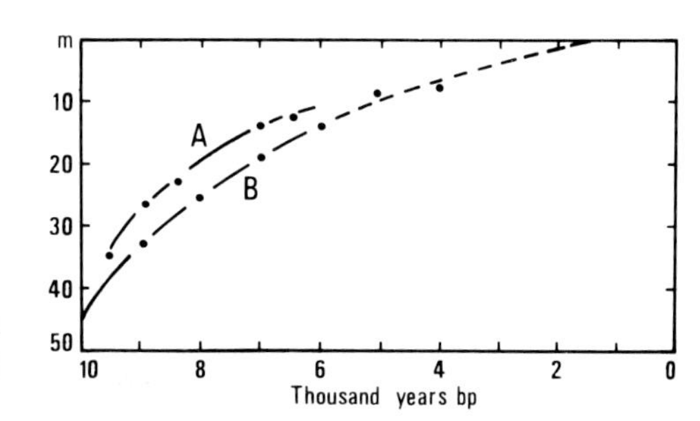

Fig. 5.3. Land uplift curves at: A. Menteith Moraine, Forth Valley (Sissons & Brooks 1971); B. Lower Strathearn (Cullingform *et al.* 1980).

(Fig. 5.3) for the Forth and Tay areas respectively. The fact that the two curves are slightly different in absolute terms does not negate their value as indicators of the rate of land uplift. In both cases the rate of land uplift decreased from 8 m/1,000 years between 9,000 and 8,000 bp to 4 m/1,000 years between 8,000 and 7,000 years. So at a time of rapid eustatic rise in sea-level the rate of isostatic rebound was slowing down and therefore the marine incursion was able to take place. The fact that the eustatic rise began to slow down after 7,000 bp but that isostatic uplift continued in Scotland resulted in the marine regression and the abandonment of the Main Postglacial Shoreline.

Because of the excellent detailed work of Jardine along the north shore of the Solway Firth and of Sissons, Smith and Cullingford and their associates in the Forth and Tay Valleys it is possible to reconstruct the environmental changes which took place in these coastal environments between 10,000 and 5,000 bp in some detail. These three areas will be examined first and although the published work on the remainder of the Scottish coastline is not so detailed some general comments can be made.

The north shore of the Solway Firth

Deposits of the main Postglacial (Holocene) marine transgression have been studied along the north shore of the Solway Firth by Jardine (1964, 1967, 1971, 1975, 1977, 1980, 1981a, 1982). These sediments occur between a few metres below O.D. and 10 m above O.D. and along much of the coastline they seldom extend more than a few tens of metres inland. However, at the head of present marine inlets such as Luce Bay, Wigtown Bay and the former embayment known as Lochar Gulf there are areas of former coastal sediments which extend inland from 1–10 km (Fig. 5.4). Jardine has identified deposits of seven different sedimentary environments: beach, gulf or open bay, estuarine (including tidal flat), lagoonal, coastal bar, coastal dune and coastal marsh. The most extensive deposits are medium to fine-grained estuarine deposits and are termed carse-deposits. The chronology of the accumulation of these deposits has been established by the radiocarbon dating of biogenic material which underlies, occurs within and overlies marine inorganic sediments (Fig. 5.5).

Between 9,000 and 8,000 bp mean relative sea-level was 2–4 m below O.D. at the eastern end of the Solway Firth but about 1 m above O.D. in Wigtown Bay

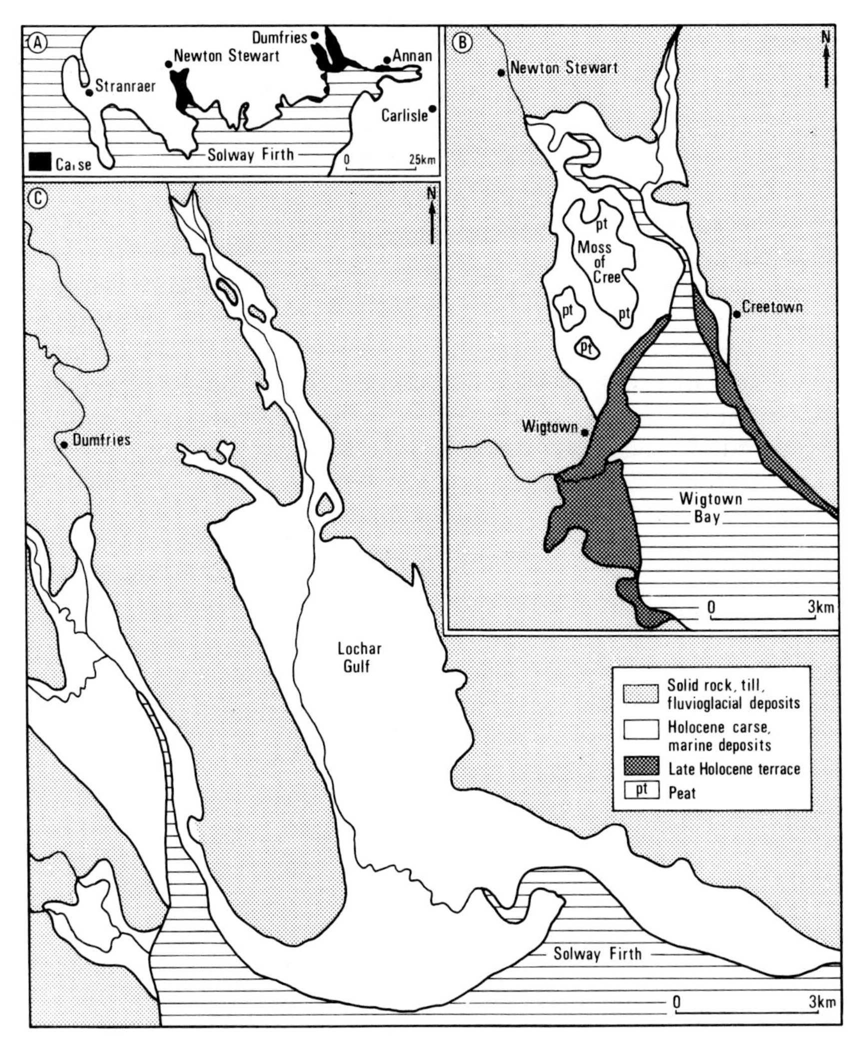

Fig. 5.4. A. North shore of Solway Firth: main embayments occupied by the sea at the maximum of the Main Holocene (Flandrian) transgression (Jardine 1977).
B. Holocene marine sediments at the head of Wigtown Bay (Jardine 1976).
C. Holocene marine sediments near Dumfries (Jardine 1975).

(Jardine 1975, Figs. 10 and 11). Between 8,000 and 7,000 bp relative sea-level was rising and Jardine (1975, p. 194) states, "In south-western Scotland, by around 7,200 bp the sea essentially had attained its maximum lateral extent for the whole of the Holocene epoch. It had not, however, attained its maximum height by that time. Radiocarbon dating and sedimentary evidence suggest that the Lochar Gulf was abandoned by the sea at approximately 6,600 bp, possibly by growth of gravel and sand bars in the vicinity of the mouth of the gulf, while contemporaneously marine conditions persisted in the eastern Solway Firth. . . . There the main Holocene marine transgression . . . did not cease until around 5,600 bp . . . and in the western part of the Solway Firth it appears to have terminated around 5,000 bp."

The work of Jardine in the Solway Firth described above is fundamentally different from that of Sissons, Smith and Cullingford and their associates in the Forth and Tay areas. Jardine has concentrated on the sedimentary record and has stressed the importance of local geomorphological and tidal conditions in controlling sediment accumulation during the marine transgression. Sissons and his associates have concentrated on the morphological and altitudinal relationships of the sedimentary accumulations and they have therefore produced more information about the assumed altitude of the main Postglacial shoreline at various places and at various times, while tending to ignore some of the implications of tidal conditions

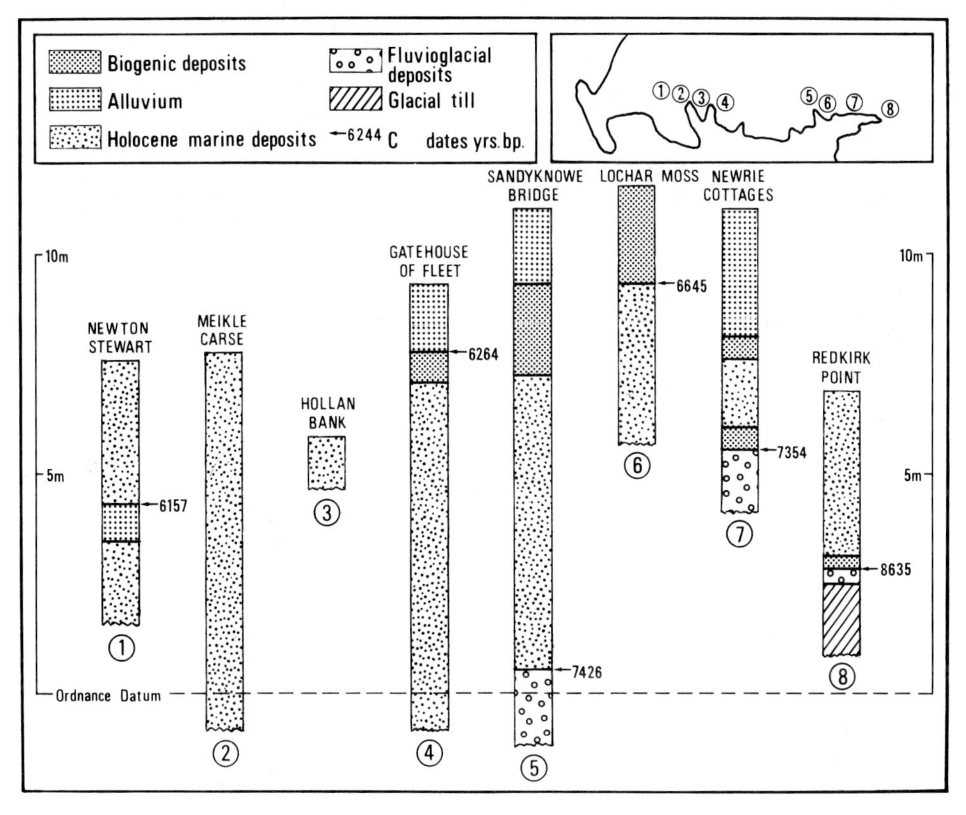

Fig. 5.5. Sedimentary successions at eight locations on the north coast of the Solway Firth (Jardine 1971).

and sedimentary environments. It is interesting to note that Sissons' isobase map (Fig. 5.1) tends to underestimate the maximum altitude of Holocene marine sediments on the north shore of the Solway Firth (9 m in the Lochar Gulf: Jardine, 1975).

The east coast—Dunbar to Fraserburgh

The publication of two papers by Sissons (1962, 1963b) on the raised shorelines of the Forth Valley was the beginning of a major re-examination of the evidence of changes in relative sea-level around the Scottish coastline. During the last 20 years Sissons has been responsible either directly, or indirectly by the supervision of numerous doctoral research projects, for the detailed investigation of the 'raised beaches' of several hundred kilometres along the east coast of Scotland. The best developed and least fragmented raised shoreline in Scotland is that of the main Postglacial (Holocene/Flandrian) marine transgression. The deposits associated with this transgression, in the relatively sheltered firths of the Forth and Tay, constitute the "carse clays" that are now raised estuarine mud flats. The junction of these carse clays with glacial drifts or solid rock constitutes the shoreline of the high stand of relative sea-level which occurred between 7,000 and 6,000 bp. The altitude of this shoreline has been accurately determined at over 1,100 points and it is known to occur at a maximum altitude of 15 m at the western end of the Forth Valley, (Sissons *et al* 1966), 10 m near Perth (Cullingford *et al* 1980), 7 m at Montrose (Smith *et al* 1980) and 2 m near Fraserburgh (Morrison *et al* 1981).

The clays, silts and sands which constitute the "carse clays" are well exposed along river banks and have been penetrated by many hundreds of boreholes. As early as 1865, Jamieson recorded peat lying beneath the carse clays of the Tay Valley and in other areas, and he concluded that the carse clays had accumulated during and following a marine transgression. Some 15 to 20 whale skeletons have been found (Sissons 1967a), associated with the carse clays, some of them many kilometres west of Stirling. The fortuitous presence of peat deposits both below and above the carse clays has allowed the chronology of both the transgression and regression to be determined by means of radiocarbon dating. Detailed work based on borehole investigations of the sediments beneath the carse clays has also revealed evidence of a phase of falling sea-level in the early part of the Postglacial period in the Forth Valley (Sissons 1966, Sissons & Brooks 1971, Smith *et al* 1978) and in lower Strathearn (Cullingford *et al* 1981).

The position of the coastline (Fig. 5.6) at the maximum of the Postglacial marine transgression in Central Scotland shows some marked changes from that of the present. Scotland was almost cut in two with only an 8 km wide land area separating the sea in the Loch Lomond basin from the sea in the Forth Valley. This change in coastal configuration was the major change in landform during the period 10,000–5,000 bp. The detailed investigations of the sediments accumulated beneath this sea allow the following reconstruction of the changes which took place during this period.

About 10,000 bp relative sea-level was falling in the upper Forth Valley. This has been revealed by the recognition of buried raised beaches to the east of the

Menteith moraine (Sissons 1966, Sissons & Brooks 1971). The High Buried raised beach (12·12 m O.D.) is thought to have been produced between 10,300 and 10,100 bp. The Main Buried Shoreline (9·9 m O.D.) has been dated only on the basis of pollen analysis at approximately 9,600 bp. The Low Buried Beach at just over 7 m O.D. has been dated at 8,690 ± 160 bp. It can be seen therefore that the chronology of sea-level changes in the upper Forth Valley between 10,000 and 8,000 bp is far from reliable but it seems reasonable to conclude that relative sea-level was falling during this period.

The age of the ensuing Main Postglacial marine transgression has been studied at a number of locations on the east coast of Scotland, where particular attention has been paid to the culmination of the event. Following Sissons & Brooks (1971), who placed the culmination at between 7,500 bp and 6,500 bp in the western Forth Valley, Smith *et al* (1980, 1982) and Morrison *et al* (1981) examined closely comparable locations further north, concluding that the culmination took place about 6,100 bp in the western Carse of Gowrie, 5,900 bp in north-east Fife, 6,700 bp near Montrose, and 5,700 bp near Fraserburgh. It is noticeable that in general the dates get progressively younger with increasing distance from the centre of uplift.

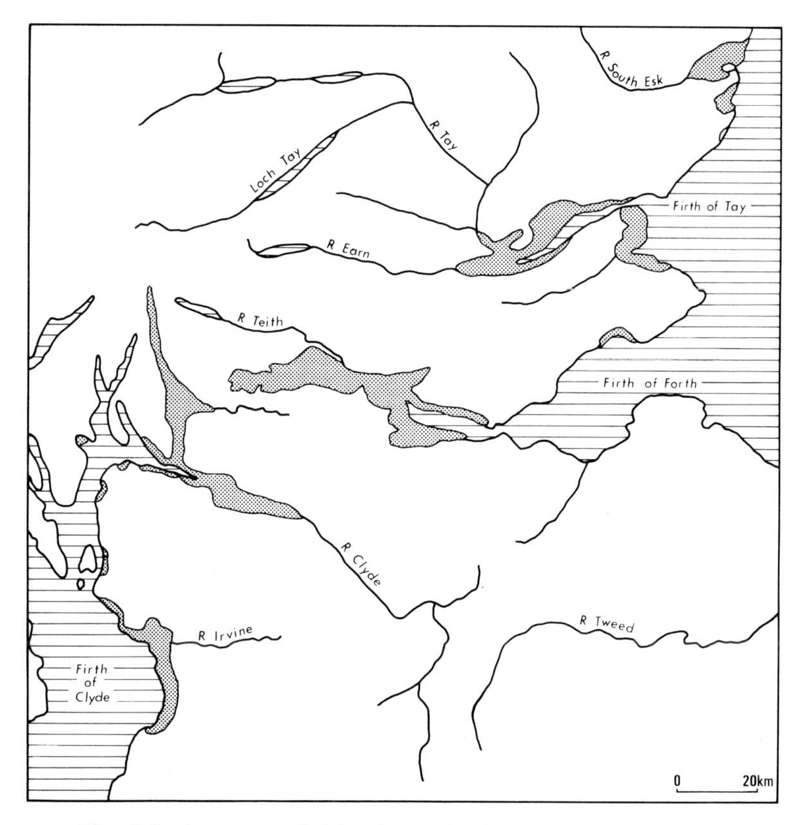

Fig. 5.6. Areas occupied by the sea in the Clyde, Forth and Tay Valleys at the maximum of the Flandrian transgression.

However, the authors point out that the peat resting on top of the carse clays may not have started to develop immediately following the withdrawal of the sea and that the dates obtained may be too young. There is also some doubt whether the carse deposits themselves (at least in the peripheral areas) relate to the culmination of the Main Postglacial transgression. It appears that at present it is not possible to determine the age of the culmination sufficiently accurately to allow the establishment of the extent of its diachroneity.

The raised shoreline of the Main Postglacial transgression is the most widespread feature related to changing sea-levels on the east coast of Scotland. The shoreline has a gradient of 0·067 m/km in the Forth Valley (Sissons 1972b). Over a distance of some 35 km the Main Postglacial Shoreline has been shown to be horizontal for 7 km in one locality and 11 km in a second locality. Sissons suggests that there was no differential uplift during the period 8,000–7,000 bp and that where gradients now exist in the Main Postglacial Shoreline they have been produced since that shoreline was produced. It seems that, "the working hypothesis that raised shorelines in areas of glacial rebound have uniform or gradually changing gradients is not of universal application" (Sissons 1972b, p. 155).

The Firth of Clyde

The Firth of Clyde has not received the detailed attention which has been given to the Firth of Forth and Firth of Tay. It is known that the Postglacial marine transgression invaded embayments at Girvan (Jardine 1967, 1971, 1975), Troon and the area around Renfrew (Fig. 5.6). The main transgression took place in southern Ayrshire sometime after 8,400 bp (Jardine 1977) but in central Ayrshire it occurred a little later—after 8,015 ± 120 bp (Welin *et al* 1975). In the Renfrew area the marine transgression began about 8,000 bp. In Ayrshire the raised marine deposits are mainly sand and they reach a maximum altitude of 12 m O.D. Around Renfrew the marine transgression cut into the Lateglacial marine deposits (Clyde beds) about 8,000 bp and the sea rose to an altitude of at least 8 m O.D. It has been suggested (Brown 1980) that extensive peat mosses prevented the penetration of the sea into much of the low ground underlain by Lateglacial marine sediments in the Renfrew-Paisley area at the time of the maximum stand of the sea during the Postglacial period.

Rose (1980) has identified Postglacial raised beach deposits (Plate 5.4) along the north shore of the Clyde estuary between Helensburgh and Dumbarton and within the Loch Lomond Basin at between 10 and 14 m O.D. A Postglacial marine incursion into the Loch Lomond basin has also been identified in a core taken from the floor of the southern basin of the Loch (Dickson *et al* 1978). Mollusca relating to the Postglacial marine transgression were recorded last century from Rossarden near Luss (Brady *et al* 1874). A layer of sediment some 70 cm thick in the Loch Lomond core is characterised by the presence of marine plankton and the absence of freshwater plants. The authors believe that the marine transgression began about 6,900 bp and ended about 5,400 bp. The fact that the marine transgression began some 1,400 years later in Loch Lomond compared with the Forth Valley is explained by the time necessary to submerge the substantial barrier across the Vale of Leven at the south-western entrance to the Loch.

Gemmell (1973) has studied the Main Postglacial Shoreline in Arran and points out that it varies considerably in altitude over quite short distances. He states that (p. 30), "In the north of the island it is generally found at altitudes between 7 and 10 m, though near Lochranza it is only 5–7 m above sea-level. In the south of the island the shoreline deposits lie at altitudes of 10–13 m. This discrepancy appears to be the result of the greater fetch available for the propagation of waves in the south of the island."

It is unfortunate that detailed comparisons of the timing and extent of the Postglacial marine transgression cannot be made between the Firth of Forth and the Firth of Clyde. It appears that the evidence from the two areas is incompatible in that the sheltered estuarine conditions of the Forth did not occur over a sufficiently extensive area in the Firth of Clyde to make detailed comparisons possible. In the Firth of Clyde it can only be concluded that the marine incursions took place sometime between 8,000 and 5,000 bp and that along some stretches of the coast the rising sea (relative to the land) simply reoccupied platforms excavated during the Lateglacial period. Deposits associated with the transgression are now found at altitudes between 8 and 14 m above O.D.

The west coast: Mull of Kintyre to Loch Broom

Studies by McCann (1966), Synge (1966) and Synge & Stephens (1966) and Gray (1974) of the Main Postglacial Shoreline along the west coast have revealed the problems associated with the development of techniques to analyse the sedimentary and morphological evidence produced by a marine transgression along an indented coastline with a complicated history. Gray (1974, p. 130) believes that "the lines constructed in these earlier studies [by McCann, Synge and Stephens] in western Scotland have only a low probability of being valid, synchronous shorelines." However, Gray's own published work at the time he wrote (1974) was limited to the Firth of Lorn and Sound of Mull and the only other published work relating to the areas further north and south was that of the other authors. Since Gray's critical comments of that work relate primarily to the techniques of measurement and the interpretation of certain coastal forms as indicators of former mean sea-levels, the major weakness in the earlier papers appears to be in the establishment of 'synchronous shorelines'. This author would argue that the possibility of establishing the synchroneity of any former shoreline fragments without the supporting evidence of dated deposits associated with those shorelines is an impossible task on such a varied coastline as the one under discussion. It is known from the excellent evidence obtained from the more sheltered areas of the Firth of Forth and Firth of Tay (see above) that the main Postglacial marine transgression probably affected much of the central west coast to altitudes of 8 to 14 m above O.D. (Fig. 5.1). Such a marine incursion would have reoccupied the Lateglacial Marine platforms and beaches which occur between 0 and 10 m O.D. throughout the same area. The great range in exposure exhibited by this coastline (e.g. compare the west coast of Jura with the Firth of Lorn) means that the landforms associated with any particular shoreline may have an altitudinal range of as much as 8–10 m (Ritchie 1966). It therefore seems most unlikely that the techniques of measurement and interpretation developed in the study of Postglacial estuarine deposits in the Solway Firth, Firth of Forth and Firth of Tay can be applied to the much more complicated West Highland coastline. Since there are no dates relating to the onset, culmination and regression of the Postglacial marine transgression on the west coast little more can be said at present than that it occurred sometime between 8,000 and 5,000 bp and that landforms and deposits associated with it can be found between 6 and 14 m above present mean sea-level (O.D.) both on the mainland coast between Kintyre and Loch Broom and on the coasts of Islay, Jura, Mull and Skye. The only areas in which the coastline at the maximum of the Postglacial marine transgression differed in position by more than several tens of metres from that of the present coastline was in embayments such as that at Crinan (Fig. 4.9).

The Northern Highlands, Outer Hebrides, Orkney and Shetland

All of these areas were peripheral to the major areas of ice accumulation during the Late Devensian ice sheet glaciation and therefore were least depressed and therefore responded least to isostatic rebound. All the evidence, although it is rather limited in amount, points to a rate of rise in sea-level which at least at times outpaced the rate of rise of land (von Weymarn 1974). In the Outer Hebrides (Ritchie 1966) it has been suggested that a continuous rise of relative sea-level took place from 13,000 to 5,700 bp, with only minor changes associated with changes in coastal configuration taking place since that time. It has also been suggested that it was this continuous rise in relative sea-level which was responsible for sweeping the great quantities of shell sand across the shallow off-shore shelf to produce the extensive machair beaches of many of the Western Isles.

Little work has been done on the northern mainland coastline. It is likely that the features associated with the Postglacial marine transgression occur at altitudes of between 0 and 4 m O.D. The occurrence of offshore peat deposits suggests that both Orkney and Shetland have experienced marine transgression throughout the Postglacial period. The existence of submerged peat, dated at 6,970 ± 100 bp (Hoppe *et al* 1965), about 9 m below present high water mark on the coast of Whalsay in the Shetland Islands indicates that marine transgression has occurred since about 6,600 bp.

Although a marine transgression produced significant changes in the coastline of mainland Scotland about 6,000 bp the extent of the invasion of the sea on to the land varied greatly depending upon the local rate of isostatic rebound and coastal configuration. In the sheltered firths of the Solway, Forth and Tay there were major indentations which produced extensive carselands when the relative sea-level fell. On the west coast only limited modification of the coastal forms produced during the Lateglacial period took place. In the Outer Hebrides and in Orkney and Shetland the coastline position has changed very little since about 8,000 bp except for the changes brought about by the migration of machair sands. Little is known about the chronology of the regression of the sea around the Scottish coast. Two or three younger shorelines have been recognised below the altitude of the highest Main Postglacial Shoreline on Mull (Gray 1972) in Galloway and Ayrshire (Jardine 1975) and in the Firth of Forth (Smith 1968) and the Firth of Tay (Cullingford 1972). The dating of the beginning of the regression has only been clearly established in the sheltered Firths but it is probably safe to say that by 6,000 bp the change to a falling relative sea-level had begun around much of Scotland's coastline apart from the Outer Hebrides and the Orkney and Shetland Islands.

In addition to the development of the raised estuarine flats (carse lands) it is very probable that the Postglacial marine transgression was largely responsible for sweeping vast quantities of sand and gravel landwards from the shallow fringing continental shelf areas (Ritchie 1972). It has been estimated that some twenty per cent of the Scottish coastline is fringed by sand accumulations in the form of dunes, links or machair. The chronology of the development of these dune systems is only poorly understood but it is likely that the prograding sea provided an impetus to their development and during the subsequent marine regression the expanding foreshores provided new sources of sand to be added to the dune systems. The carse lands, dune systems and raised beaches all played a significant role in the early settlement patterns developed by the Mesolithic and Neolithic populations of Scotland.

VEGETATION AND SOILS

The Postglacial history of the evolution of the vegetation of Scotland has been deduced either from macroscopic plant remains in peat bogs or from both macroscopic remains and microscopic pollen grains and spores contained either in peat or lake sediments. The early studies of Lewis (1905, 1906, 1907, 1911) and Samuelsson (1910) were followed in 1923 by the first pollen diagrams from Scottish peat and lacustrine deposits (Erdtman 1923, 1924). Over the next thirty years less than a dozen papers based on palynology were produced (Edwards 1975). A major contribution to the history of Scottish woodlands was made by Durno (1956, 1957, 1958, 1959) and he demonstrated that the forests in all parts of Scotland were most widespread in the Boreal and Atlantic periods (9,000–5,000 bp) and declined to their lowest levels in the Sub-atlantic period (after 2,500 bp).

With the application of radiocarbon dating to profiles studied by palynologists the rate of colonisation and subsequent changes in the forest cover were established (Moar 1969a, b, Nichols 1967, Birks H. H. 1970, 1973, Birks H. J. B. 1973b, Pennington *et al* 1972, Hibbert & Switsur 1976, Birks & Matthews 1978, O'Sullivan 1976, Peglar 1979, Dickson *et al* 1978). These studies have revealed that much of Scotland except for Caithness, Orkney, Shetland and the smaller Hebridean islands was rapidly colonised by forests at the beginning of the Postglacial period. Birch was ubiquitous in both highland and lowland Scotland. Pine was more common in the north-east while oak was most significant in the south and west. H. J. B. Birks (1977) has summarised the forest history of Scotland in the context of the four major potential vegetation regions identified by McVean & Ratcliffe (1962). The native Scottish forest formed a western phase of the European temperate deciduous forest-coniferous forest-boreal deciduous forest transition but only small parts of the original forest cover remain today (Fig. 5.7). They consist primarily of oak (*Quercus petraea, Quercus robur*), pine (*Pinus sylvestris*) and birch (*Betula pubescens* and *Betula*

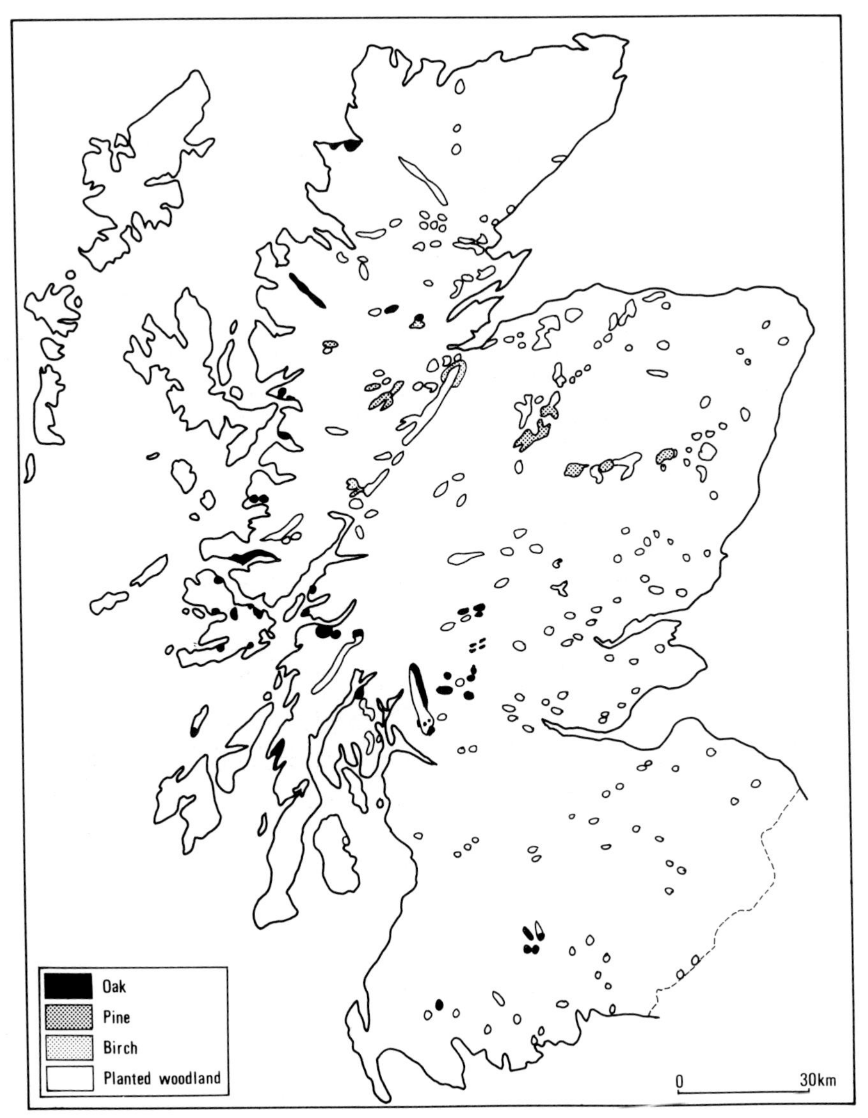

Fig. 5.7. Present distribution of Oak, Pine and Birch woodland (McVean & Ratcliffe 1962). Planted woodland, mainly coniferous, is shown in otuline. Reproduce with the permission of the Controller of Her Majesty's Stationery Office.

verrucosa). McVean and Ratcliffe also attempted to reconstruct the distribution of these woodland types during the present climatic period but prior to the onset of large scale human forest clearance (Fig. 5.8). This map was based on the available evidence provided by the present distribution of woodland types and tree species

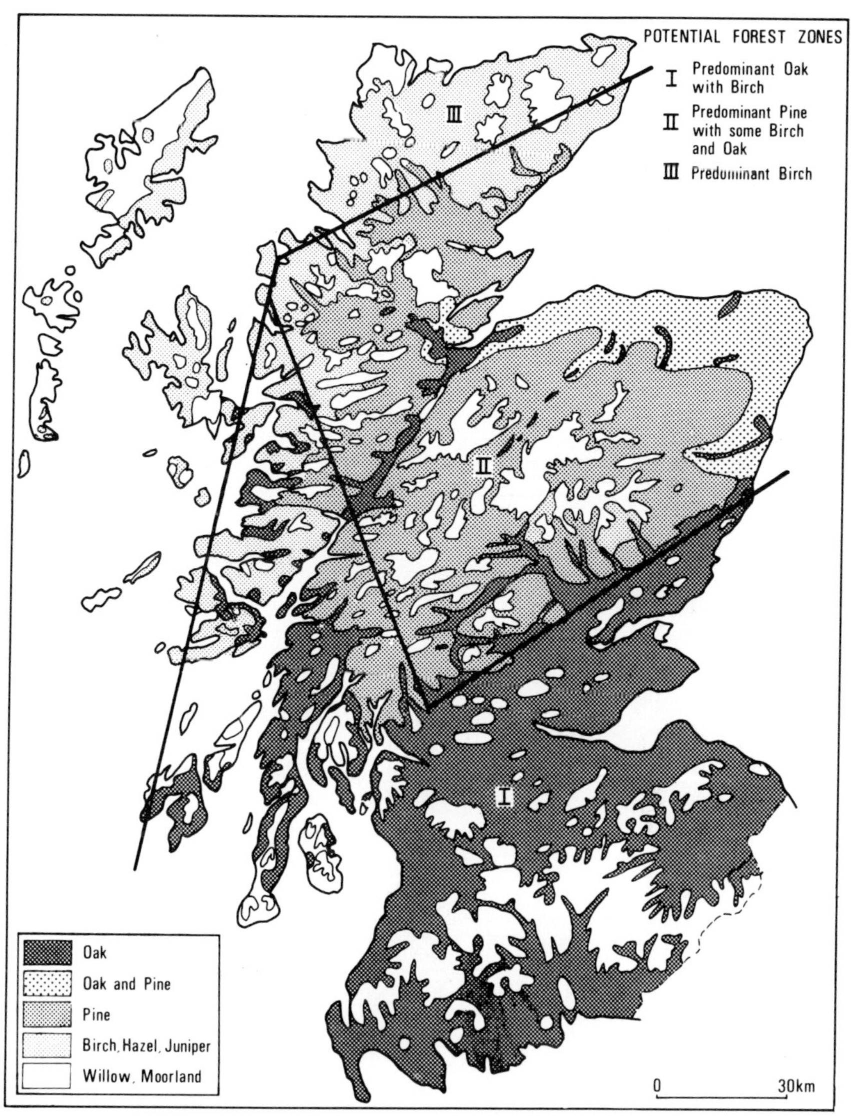

Fig. 5.8. Reconstructed distribution of Oak, Pine and Birch woodland during the present climatic period but before the onset of large-scale human forest clearance (McVean & Ratcliffe 1962). Reproduced with the permission of the Controller of Her Majesty's Stationery Office.

and their known ecological requirements along with the then available information provided by pollen analysis of, and macroscopic remains in, peat bogs. McVean & Ratcliffe point out (p. 11), ". . .while a simple climatic zonation of forest types is not found in the Highlands [they stress the importance of edaphic controls], some climatic control is discernible, particularly in latitudinal distribution." The four major potential vegetation regions identified by McVean and Ratcliffe will be used

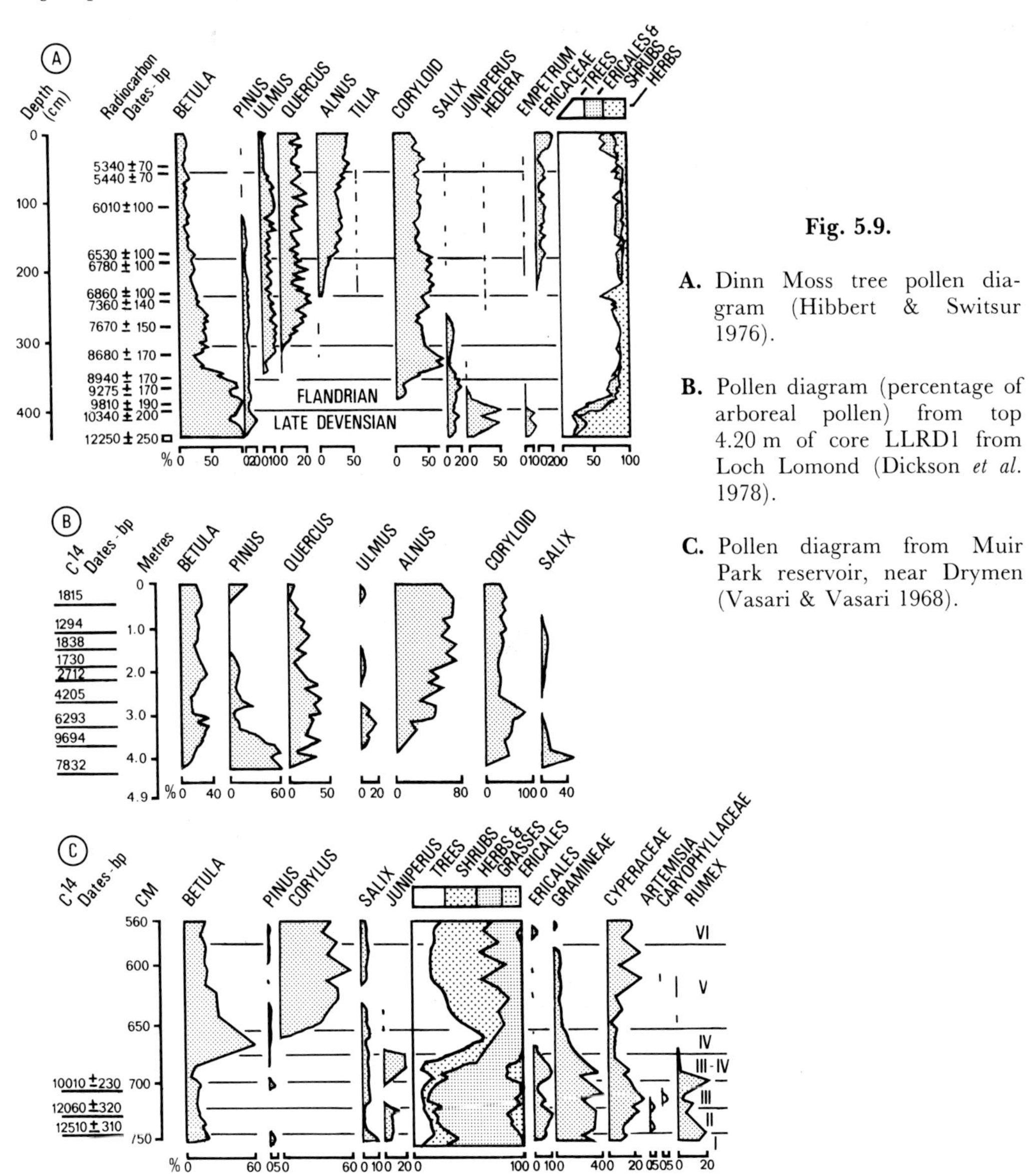

Fig. 5.9.

A. Dinn Moss tree pollen diagram (Hibbert & Switsur 1976).

B. Pollen diagram (percentage of arboreal pollen) from top 4.20 m of core LLRD1 from Loch Lomond (Dickson *et al.* 1978).

C. Pollen diagram from Muir Park reservoir, near Drymen (Vasari & Vasari 1968).

to describe the vegetational history of Scotland both in this chapter and the next. The regions can be described as follows (Fig. 5.8):

I *Oak forest with birch*. This region covers the whole of southern Scotland south of the Grampians and includes all of Argyll, Arran, Jura, much of Islay and Mull and the area between Loch Linnhe and the Sound of Sleat.

II *Pine forest with some birch and oak*. This region contains the remainder of the Highlands and north-eastern lowlands.

III *Birch forest*. This region covers the extreme northern part of the mainland north of a line from Ullapool to Wick and much of Skye and the Outer Hebrides.

IV *Grasslands and heaths*. This is a treeless region covering Orkney, Shetland and the smaller inner Hebridean islands.

The general vegetation history of Scotland has been known for some years and in particular the destruction of much of the forest cover by man has been appreciated. However, during the last decade numerous palynological investigations have included closely spaced radiocarbon dates to identify the pollen zone boundaries. In the regional descriptions which follow, heavy reliance will be placed on these studies while at the same time relating the less well-dated pollen diagrams to them.

1. The oak-birch forest region

Investigations by Nichols (1967), Moar (1969a) and H. H. Birks (1972a) in south-west Scotland, Hibbert & Switsur (1976) and Mannion (1978) in south-east Scotland, Newey (1965) in the eastern central lowlands, Vasari & Vasari (1968) and Dickson *et al* (1978) in the Loch Lomond area, and Nicols (1967) and Rymer (1974) in Argyll provide a good basis for the interpretation of vegetation colonisation and development during the first 5,000 years of the Postglacial period. Throughout the region birch (*Betula*) and hazel (*Corylus*) soon replaced the open treeless vegetation of the Lateglacial Stadial. At Din Moss in Roxburghshire (Hibbert & Switsur 1976) birch and hazel were established by 9,700 bp (Fig. 5.9a) and within a further thousand years a mixed deciduous forest of oak, elm and hazel had developed. Some pine did occur but was not very significant except in marginal habitats such as dried peat bog surfaces in Galloway between about 7,500 and 6,800 bp where the pine stumps were subsequently buried by peat growth (Birks 1975). Alder (*Alnus*) replaced willow (Salix) in wet habitats between 7,000 and 6,500 bp.

In the Loch Lomond area (Fig. 5.9b, c), birch was again the initial coloniser but it was soon joined by hazel and a mixed woodland of birch, oak, elm and hazel was well established by 6,500 bp. There was a rapid increase in the amount of alder present just after 6,000 bp. A similar vegetation history has been identified for the Forth Valley and the Lothians by Newey (1965, 1967).

Over much of the area covered by the mixed deciduous forest, the soils developed profiles which were in equilibrium with the vegetation and the climate and therefore over extensive areas brown forest soils developed. In the west and south-west of the country, where acid soils had developed the existence of a forest cover meant that strong podsolisation did not occur.

About 7,500 bp, the beginning of the Atlantic period (Pollen zone VIIa), the climate of Scotland is believed to have become somewhat wetter and warmer. At this time deciduous trees reached altitudes of over 800 m in central Scotland, (e.g. on Ben Lawers in Perthshire). However, in this wet and mild climate great areas of sphagnum peat began to develop on former lake floors, and on ill-drained flat areas. The continued growth of the peat eventually rose above the ground-water table and it became ombrogenous, i.e. the peat forming vegetation became separated from ground water and mineral soil and became dependent on rain water alone for its supply of mineral salts and therefore the peat became very acid. In such circumstances only a specialised vegetation will survive (*Sphagnum* species). In areas of very heavy rainfall ombrogenous peat even developed on sloping ground to produce blanket bogs. At present some 10 per cent of Scotland is covered by peat (over 30 cm thick) and many of these peat bogs originated over 6,000 years ago.

The expansion of peat bogs towards the end of this period must have reduced the extent of the oak-birch forest but it was the arrival of man and his associated attempts at woodland clearance which had the greatest impact on the extent of the forest. The so-called "elm decline" will be discussed in greater detail in Chapter 6, but it has been recognised in many pollen diagrams and dated as beginning at 5,080 ± 100 bp in the Galloway Hills (H. H. Birks 1972a), at 5,440 ± 70 bp in Roxburghshire, Din Moss (Hibbert & Switsur 1976), at 5014 ± 120 bp in the Forth Valley (Godwin and Willis 1959) and at 5,300 bp in Loch Lomond (Dickson *et al* 1978). After 5,000 years of natural forest development it seems likely that it was the impact of man on the environment rather than any major climatic change which was to result in the first significant change in environmental conditions.

2. The pine forest (with some birch and oak)

Detailed work in the North-west Highlands (Pennington *et al* 1972), in the Spey Valley (H. H. Birks 1970, H. H. Birks & Matthewes 1978, O'Sullivan 1976) and in the Dee Valley (Edwards 1978, Edwards & Rowntree 1980) enable the development of the Scottish pine forest to be described with some confidence. Although there are certain local variations between the various sites for which radiocarbon dated pollen profiles are available the following general pattern of vegetation history can be deduced (Fig. 5.10).

At the beginning of the Postglacial period birch began to colonise the *Juniper-Empetrum* association. Soon after 9,000 bp hazel became an important element in the vegetation and by about 8,000 bp the whole of this region was covered by a birch-hazel forest except that in the Great Glen, elm and oak were present. About 8,000 bp pine began to invade the birch-hazel forest and became dominant in the Abernethy Forest by 7,200 bp and in the north-west Highlands by 6,500 bp. After 6,500 bp there was a significant increase in alder in all parts of the region and at about 5,300 bp the elm decline began.

Pennington *et al* (1972) note that there was a sudden expansion of pine just after 8,000 bp at Loch Sionascaig with a corresponding decline of birch, hazel and willow. A rapid expansion of pine took place in the Abernethy Forest about 7,300 bp (Birks & Matthews 1978). It may well be that throughout much of the pine forest region early acidfication of soils, a general delay in the arrival of deciduous trees and the general poverty of the underlying rocks and drifts, produced areas of soil which could never thereafter have been expected to support either hazel or mixed-oak forest. However, the Great Glen did develop a mixed-oak forest (birch, hazel, elm and oak) as early as 8,000 bp and this was invaded by some pine and alder about 6,000 bp.

The site investigated by Pennington *et al* (1972) at Loch Sionascaig (Fig. 10A)

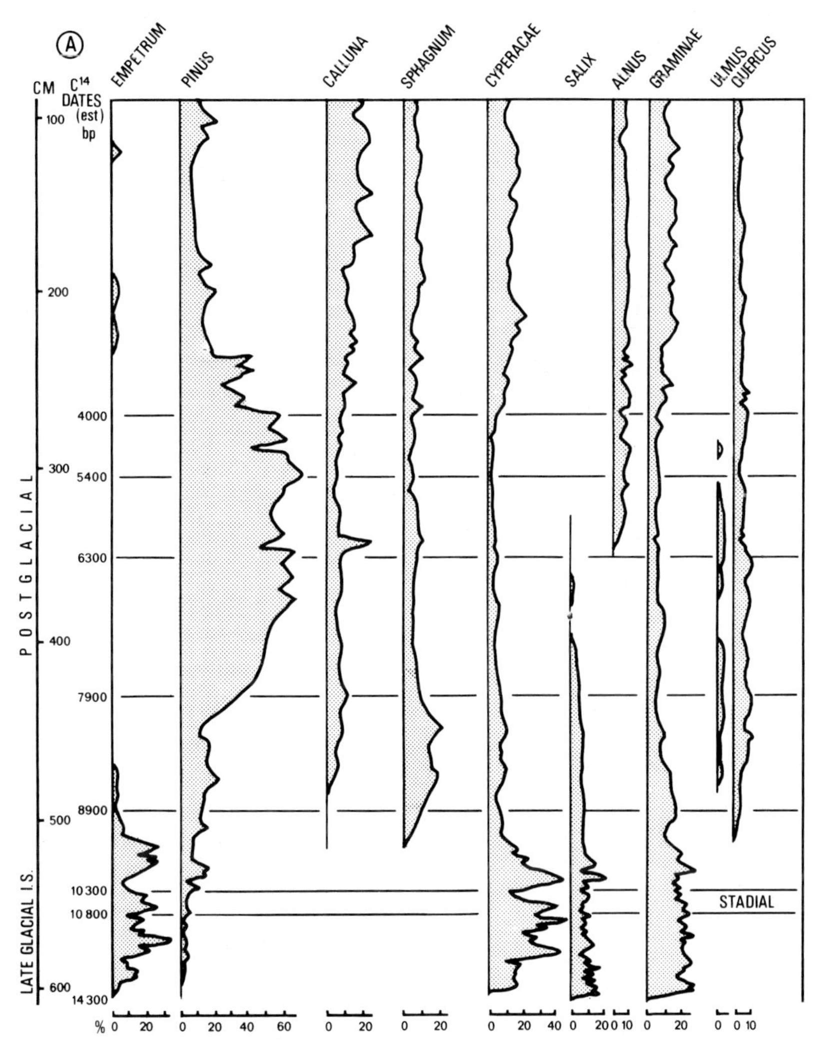

Fig. 5.10. A. Percentage pollen diagram from Loch Sionascaig (Pennington *et al.* 1972).

would suggest that the boundary between the pine-birch forest region and the birch forest region drawn by McVean and Ratcliffe (Fig. 5.8) should be at least some 15 km further to the north-west. However, such minor modifications to the regional forest boundaries are to be expected as more palynological information becomes available.

There appears to have been a marked increase in the rate of formation of

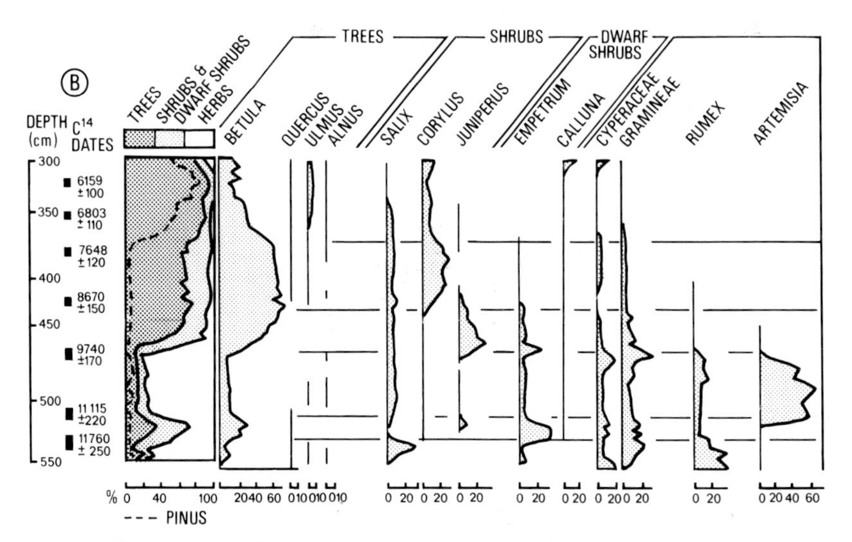

Fig. 5.10. B. Percentage pollen diagram (selected taxa) from Abernethy Forest, Inverness-shire (Birks & Matthewes 1978).

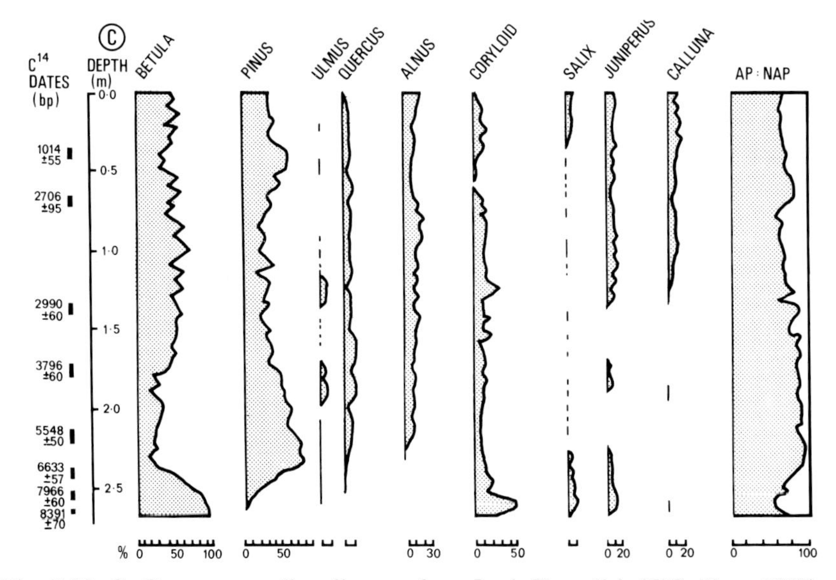

Fig. 5.10. C. Percentage pollen diagram from Loch Pityoulish (O'Sullivan 1976).

blanket peat in the north-west Highlands about 5,000 bp. At Loch Claire, however, there is no evidence for any expansion of peat growth on the steeper slopes surrounding that site but there is evidence of both forest fires and the presence of man at an horizon dated to 5,360 bp. This appears to be the earliest trace of human influence on vegetation in the north-west Highlands. On Speyside human effects on the pine/birch forest did not occur until 3,600 bp (O'Sullivan 1976).

The pollen evidence clearly shows the dominance of pine both in the Cairngorms, on Speyside (Fig. 5.10B, C) and in the area to the north-west of the Great Glen from about 7,000 bp to 5,500 bp. However, there seems to have been some regional variation in the relative proportions of different tree types (particularly birch and pine) and in the timing of the expansion of pine in the west (8,300 bp)

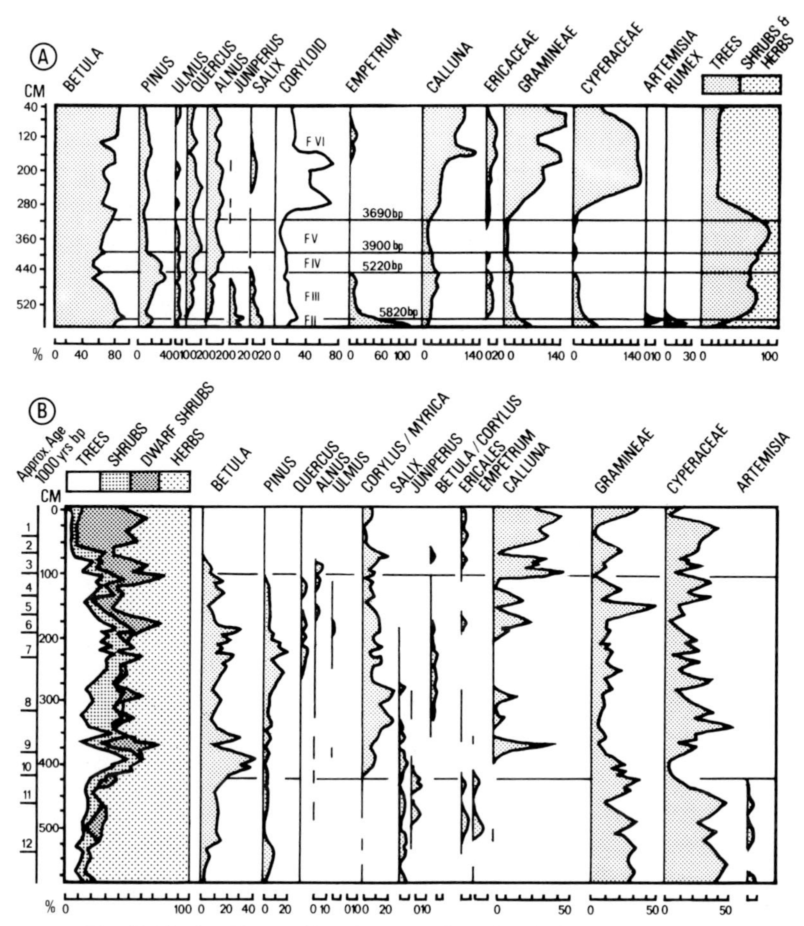

Fig. 5.11. A. Tree pollen diagram from Loch Duartbeg, Sutherland (Moar 1969).
B. Percentage pollen diagram from Loch of Winless, Caithness (Peglar 1979).

and in the east (7,000 bp). It is probable that the oldest pine forests in Scotland are in Wester Ross rather than in the Cairngorms. The altitudinal limit of the forest is not known but pine and birch fragments have been found in blanket peats up to 790 m in the Cairngorms (Pears 1968, 1972). Above the pine and birch there was probably a belt of Juniper scrub succeeded by dwarf shrub and heath at higher altitudes.

3. The birch forest

Northern Sutherland and Caithness and southern Skye (Moar 1969b, Peglar 1979, H. J. B. Birks 1973a, H. J. B. Birks 1977) developed birch and hazel woodlands by about 9,000 bp, (Fig. 5.11). In southern Skye the birch-hazel woodland was dominant from about 9,600 to 6,500 bp. Alder became an important local component at about 6,500 bp but oak and pine were never very significant. In the more exposed basaltic areas of northern Skye the available pollen profiles indicate that a closed forest never developed with trees (birch, hazel and poplar) only growing in sheltered localities. A grass and tall-herb association developed on the moist basalt lowlands. Alder expanded at about 6,500 bp but was never abundant. In northern Sutherland the birch-hazel woodlands were well-established by 8,500 bp and apart from an expansion in Alder about 6,500 bp they remained dominant until after 5,000 bp. Elm was present on the Durness limestone.

In Caithness, the birch-hazel expansion occurred at about 9,300 bp but only formed patches of woodland in sheltered localities. Willow and tall-herb and fern dominated communities were widespread and the pollen evidence indicates that a closed forest never developed in Caithness.

4. Grasslands and heaths

The Orkney and Shetland Islands (H. J. B. Birks 1977, Keatinge and Dickson 1978) never developed a forest cover. The early and middle Flandrian vegetation of these islands consisted of tall herbs and ferns with occasional patches of birch and hazel scrub. No evidence has been found of pine, oak and alder or other major forest trees as indigenous trees. The vegetational history of Harris and Lewis is not known but the smaller islands of Barra and Canna (Blackburn 1946, Flenley & Pearson 1967) never supported a forest cover but maritime grasslands and dwarf-shrub heaths have existed since the beginning of the Postglacial period.

FAUNA

Surprisingly little is known about the early Postglacial animal migrations into Scotland (Ritchie 1920, MacGregor & MacGregor 1936 and Lacaille 1954). Most authors are agreed that the early Postglacial faunas reflect a rapid climatic change with thermophilous assemblages rapidly establishing themselves. The characteristic large mammals of the period prior to 27,000 bp did not reappear and the Holocene fauna was generally very similar to that of today except for certain notable

Postglacial extinctions. This is particularly true of both the marine and non-marine molluscs.

The animal populations associated with the spread of birch, pine and oak forests in Scotland between 9,000 and 6,000 bp included the following species: Red deer (*Carvus elaphus linn.*), giant fallow deer (*Dama dama*), great elk (*Alces alces*), wild horse (*Equus caballus ferus nel*), great ox (*Bos primigenius boj*), beaver (*Castor fiber linn*), wild boar (*Sus scrofa ferus linn*), wolf (*Canis lupus*), brown bear (*Ursus arctos linn*), lynx (*Lynx lynx linn*), and variable hare (*Lepus variabilis linn*).

There is no information available about the relative importance of these various species in Scotland or about their geographical distribution. The animal population was presumably at its greatest in the generally forested landscape of about 6,000 bp. Presumably it was the arrival (for the first time as far as we know) of *Homo sapiens* around 8,000 bp which constituted the greatest threat to the established animal population rather than any change in the natural environment.

CLIMATE

Our knowledge of Scotland's climate during the early Postglacial period is based almost entirely on the general understanding of the relationship between plants and temperature conditions on a continental scale (Lamb 1972). Changes in the vegetation of north-west Europe during the period have been determined by the

TABLE 5.1

Thousand years bp	Pollen zones	Climatic type	Mean July Temp °C	Vegetation	Sea level (m) Ardyne	Cultures
5			16	Elm decline		
6		Climatic Optimum		Oak/Pine	+13	
7	VIIa	Atlantic				
8	VI					Mesolithic
9	V	Boreal	15	Hazel/Birch/Pine		
10	IV	Preboreal		Birch		
11	III Loch Lomond Stadial	Sub Arctic	7–9	Tundra	+10	

study of pollen and macro plant remains in peat and lake sediments (Blytt 1876, Sernander 1908, Godwin 1975). The establishment of the main vegetation patterns at various stages in the Postglacial period led to deductions about the main climatic conditions prevailing at each stage. Significant changes in vegetation/climate have been further clarified by studies of cores of sediment from the ocean floors involving the analysis of their foraminiferal content. Oxygen isotope analysis of these ocean cores and of cores from the Greenland ice sheet have also added to the general understanding of climatic conditions in the northern hemisphere throughout the Postglacial period.

The pattern of vegetation change between 10,000 and 5,000 bp in Scotland described in the previous section fits into the more general pattern identified for north-west Europe as a whole and it seems reasonable to interpret the Scottish vegetational sequence in the same general terms of climatic conditions as have been identified for the much larger area. Although the pollen zones and the Blytt and Sernander climatic zones are only broadly synchronous the general sequence of environmental changes forms a useful framework.

The period 10,000 to 5,000 bp (Table 5.1) includes four pollen zones (IV, V, VI, VIIa) and three climatic types (Pre-Boreal, Boreal, Atlantic). There is no doubt that the transition from the grass-heath and shrub vegetation of the Lateglacial stadial to a forest cover (birch-hazel-oak-pine) represents a marked improvement in Scotland's climate over a relatively short period and most of the evidence points to a climate very similar to that of the present day being established by 9,000 bp. Throughout the Postglacial period there is no evidence which suggests any marked fluctuations in temperatures in Scotland (i.e. more than 5°C in mean monthly average temperatures), similar to those which occurred during the Lateglacial period. The main problem is to identify the relatively small temperature and precipitation fluctuations which may have occurred and which might account for significant changes in vegetation patterns.

On the basis of a wide range of evidence Lamb (1972) has suggested probable temperature and precipitation conditions expressed as departures from present (last 100 years) averages (Table 5.2). He also points out that the quoted figures are broad generalisations and should probably be regarded as 1,000 year averages. No

TABLE 5.2

	Pre Boreal 10,000–9,000 bp	Boreal 9,000–7,000 bp	Atlantic 7,000–5,000 bp
Temperature	Rapid rise	Summer: $+\frac{1}{2}$°C Winter: −1°C	Summer: +2°C Winter: +1°C
Precipitation	5–8% less		10–15% more
Evaporation	2–6% less		8–14% more
Discharge	6–11% less		12–16% more

attempt is made to identify local variations in Scotland produced by aspect, altitude or latitude.

It might be argued that monthly temperature variations of 1° or 2° centigrade over a thousand years are far less significant than short term but marked deteriorations which might be hidden in the long term averages. It must also be stressed that the precipitation, evaporation and discharge figures are probably very unreliable in terms of any particular small area. The presence of a forest cover almost certainly increased evapotranspiration and decreased discharge in forested river basins but it is possible that these trends were counterblanced by increased precipitation totals. It can certainly be argued that Scotland probably enjoyed a more maritime climatic environment at the time of the Main Postglacial marine transgression (circa 6,000 bp) as a result of the large bodies of water which occupied the Forth and Tay Valleys and to a lesser extent the Clyde estuary (Fig. 5.6). It is also possible that during the Atlantic period, sometimes referred to as the climatic optimum, the altitudinal contrasts in Scotland's environment may have been less marked than now. It is probable that the tree-line reached higher altitudes about 6,000 bp than at any other time in the Postglacial period.

Our knowledge of the regional vegetation of Scotland and of sea-level changes between 10,000 and 5,000 bp is much greater than that of regional climate. The present day regional variations of climate are quite significant and it can only be presumed that there were significant regional variations during the early Postglacial period. However, the environments encountered by the first human colonists were certainly not the severe conditions of the Lateglacial Stadial. On the basis of present evidence much of Scotland was already covered by forest (birch, hazel, oak and pine) and relative sea-level was beginning to rise rapidly when the first colonists arrived. It is also probable that temperature and precipitation conditions were not significantly different from those of the present day.

THE INFLUENCE OF EARLY MAN

The earliest evidence of man's occupation of Scotland is limited in amount and somewhat controversial in its interpretation. This evidence has recently been summarized by Morrison (1980) and it mainly consists of finds of flint and bone implements and post-holes, hearths and shell middens indicative of at least temporary settlement. Virtually all the sites so far discovered have a coastal location (Fig. 5.12) but many of them either only consist of surface finds or were excavated many years ago before the development of modern archaeological techniques. Only five sites have been radiocarbon dated: Lussa on Jura (Mercer 1968, 1970a, b, 1971, 1972, 1974a, b, c), Morton, Fife (Coles 1971), Inveravon in the Forth Valley (Mackie 1972), Barsalloch, Wigtownshire (Cormack & Coles 1968, Cormack 1970) and Oronsay (Mellars 1979, Mellars & Payne 1971). Even though there may be some debate about the interpretation of the radiocarbon dates from these sites in terms of their possible contamination and the events which they may or may not relate to, it has at least been established that the first human occupation of Scotland took place sometime between 8,000 and 6,000 bp.

Most of mainland Scotland at the time the settlement sites referred to above were occupied was covered by a closed forest and around much of the coastline a marine transgression was in progress. The majority of the sites were very close to, or on, the beaches formed by this transgressive sea or in caves in old abandoned marine cliffs. The flint and bone implements, shells, bones and the macro- and micro-vegetation evidence associated with the sites indicate that their occupants were Mesolithic hunters and gatherers. At some of the sites post-holes and hearths have also been recognised and it has been suggested that both semipermanent as well as seasonal occupation may have occurred. Morrison (1980, p. 118) states,

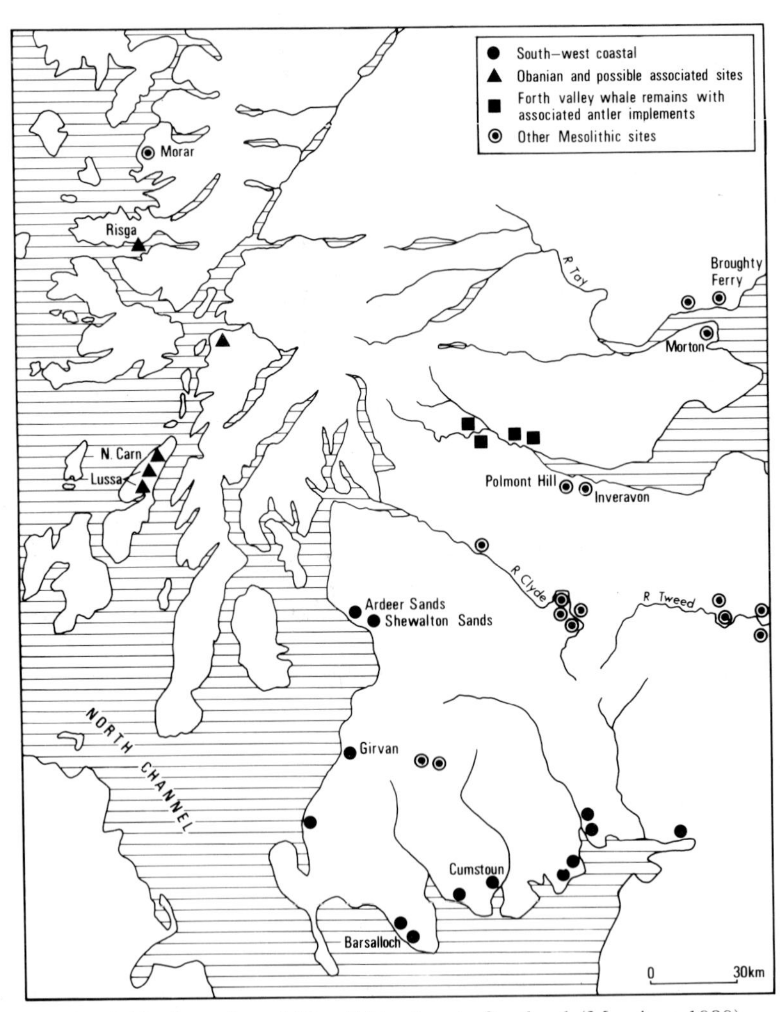

Fig. 5.12. Location of Mesolithic sites in Scotland (Morrison 1980).

"There can be little doubt that man's activities at the hunter/gatherer stage of cultural development were to a great extent controlled by the environment. The vast forests of early Holocene times were barriers to movement and dwelling sites were restricted to their fringes, or to lakeside, riverside or coastal locations. . . . But with tools, fire and an intimate knowledge of the flora and fauna within his territorial range, he was already learning to manipulate, if only on a small scale, some elements of the habitat which determined his way of life."

Although so little is known about the life style and numbers of the population of the Mesolithic period it is necessary to enquire as to the extent of the impact of these people on the natural environment. In Britain as a whole Clark (1956) suggested that the Mesolithic population was 4,500 and Piggot (1965) suggested a figure of 10,000 at a date of 9,500 bp. Atkinson (1962) on the basis of about 100 known Mesolithic sites suggested that Scotland had a Mesolithic population of 60–70. It must be concluded at present, that the numbers of people occupying Scotland between 8,000–5,000 bp are simply not known. It has been argued on the one hand that the apparent 'coastal' distribution of Mesolithic population reflected the general inhospitableness of much of the interior which would have been heavily forested but on the other hand the coastal bias may simply reflect the great facility for discovering and collecting Mesolithic tools in beach and sand dune areas.

Even if the Mesolithic population was small, man's selective use of certain types of vegetation for food, fuel or building materials along with his use of fire, whether accidental or deliberate would have had some effect on the vegetation in those areas he occupied. Unlike the numerous sites to be discussed in Chapter 6 at which Neolithic farming activities are recorded in the pollen diagrams from those sites, there is as yet no substantive pollen evidence to suggest major modification of the vegetation of Scotland by Mesolithic people. However, in Yorkshire, pollen analyses have revealed that areas of thick forest cover changed to more open conditions at levels containing charcoal and Mesolithic flints (Dimbleby 1961, 1962, Simmons & Cundill 1974, Jones 1976). In Denmark (Troels-Smith 1960) bundles of hazel and willow twigs have been discovered at a Mesolithic site and it was suggested that they may have been used for making fish traps. It has also been suggested that controlled burning of woodland by man led to the creation of a habitat suitable for the expansion of hazel and there is no doubt that it was a species used by man and eaten by the animals he hunted (Mellars 1976). Although a marked rise in hazel pollen is a widespread occurrence in the early Postglacial period, it seems unlikely that on the present limited evidence of population density that such a widespread and synchronous vegetation change could be initiated solely by human activity.

Albeit on limited evidence, it can be suggested that Scotland's Mesolithic population was dominantly coastal in location and made use of, rather than dominated or altered, the natural environment to any great extent. The population took advantage of an even greater indented coastline than exists at present and probably enjoyed a moister and slightly warmer climate than that of to-day (Table 5.2). It is possible that some of the earlier occupation sites, if they existed, were

overwhelmed by the Postglacial marine transgression. The first 5,000 years of the Postglacial period saw the development of a truly temperate environment with a dominantly forested landscape for the first time in some 30,000 years. This was the environment that attracted further human occupation which, in turn, led to its modification.

6

THE LAST FIVE THOUSAND YEARS
circa 5000 bp to the present

About five thousand years ago Scotland's environment consisted of a climate only marginally different from that of to-day, landforms inherited from geological events which took place tens of thousands or even millions of years previously and a vegetation cover which had developed over the previous 8,000 years from a grass-heathland association into an oak or pine-birch forest. The changes in the environment described in the earlier chapters of this book have been difficult to decipher because of the limited evidence available and the difficulties inherent in using that evidence in reconstructing palaeoenvironments. However, it is often easier to identify dramatic environmental changes such as those of the Lateglacial period which involved temperature fluctuations of over 5°C which in turn led to major changes in geomorphological processes and vegetation patterns, than it is to identify the minor changes which have occurred in Scotland's climate over the past 5,000 years. It is unlikely that Scotland's average winter and summer temperatures at sea level have fluctuated by more than 1·5°C and that precipitation totals have changed by more than ten per cent over this period. In terms of major changes in the nature of, and efficiency of, geomorphological processes such fluctuations are insignificant. Identifying such minor fluctuations would be difficult enough without the possibility that some of the environmental changes were either initiated, or at least assisted by human activity. Undoubtedly the most significant change in Scotland's environment during the last 5,000 years has been the transformation of the vegetation from a dominantly forest cover to a grassland-heathland association plus agricultural and urban landuse. The evidence used to identify the vegetational changes is still controversial and will be discussed in detail later in this chapter. There is little doubt, however, that the development of the Neolithic, Bronze Age and Iron Age cultures between 5,000 and 2,000 bp produced major human impacts on the landscape. It is possible that minor temperature and precipitation changes during this period caused changes in the forest cover and the expansion of peat and heath but it is also probable that the agricultural practices of the expanding population resulted in the degeneration and/or removal of the forest cover.

The availability of documentary evidence for the period since 1100 AD and of an instrumental record for the last 200 years would suggest that changes in the environment over the last 1,000 years would be much easier to identify. Although studies of documentary sources and instrumental records have been undertaken in England and on the European continent (Lamb 1977), apart from the excellent detailed work of Parry (1975, 1976, 1978) in south-east Scotland, little work has been done on this period in Scotland. A major difficulty in using documentary sources is, of course, the problem of recognising the natural changes in the environment as opposed to those induced by human activity related to a given level of technology or social organisation. It is not the intention of the writer to discuss the possible significance of relatively small environmental changes on the economic, political and social fortunes of the Scottish people. In the context of this book the environmental changes of the last 1,000 years have been insignificant but they have been shown by Parry (1978) to have been very important to agriculturalists operating in what were already marginal environments.

LANDFORMS AND SEA-LEVELS

A five thousand year time span is relatively short for geomorphological processes to bring about major changes in landforms in a temperate environment. Little is known about rates of erosion in Scotland over this time span but it can be assumed that weathering, mass movement and fluvial erosion, transportation and deposition continued to modify the hill-side slopes and individual depositional landforms which emerged from beneath the last ice sheet and the valley glaciers of the Loch Lomond Advance. Around much of the Scottish coastline, except in Orkney, Shetland and the Outer Hebrides relative sea-level was falling from the maximum altitude attained by the Main Postglacial transgression about 6,000 bp. Along much of the coastline relative sea-level has fallen between six and twelve metres since that time and minor still stands have allowed shorelines higher than the present one to form. Gray (1972) has recognised two definite shorelines in the Oban area, below the Main Postglacial shoreline, but their ages are unknown. The chronology of the emergence of much of the Scottish coastline is simply not known but the effects of this emergence have been both dramatic and very significant in human terms.

The Main Postglacial transgression resulted in large areas of the Solway, Clyde, Forth, Tay and Cromarty Firths becoming sites of estuarine deposition (carse clays). When relative sea-level began to drop these new estuarine deposits first of all became marshlands with extensive reed beds and then eventually were colonised by grasses, sedges, mosses and woodland. The raised beaches along more exposed sections of the coast, where sand and gravel, rather than silt and clay had accumulated during the Main Postglacial transgression, were also abandoned to leave benches 50–500 m wide consisting of low angle slopes underlain by well-drained unconsolidated materials upon which relatively easily worked soils began to develop. (Pl. 6.1) The creation of these new 'flat lands' around the periphery of the rugged upland which dominates so much of Scotland, was probably the most important environmental change to take place in Scotland between 5,000 and 2,000 bp. The extent of these new land areas amounted to a total of about 200,000 hectares (C. A. Halstead, pers. comm.). These new lands provided relatively attractive occupational sites for the Neolithic and Bronze Age peoples and major routeways were opened up along their margins. This very basic environmental attribute is still reflected in the modern settlement and communication patterns of Scotland.

The relative fall in sea-level around much of the Scottish coastline over the last 5,000 years has also led to the continued incision of rivers into their valley floors. Again, the details of terrace and flood-plain development in the valleys of Scotland are not known. Large quantities of gravel, sand and silt deposited in the valley systems during the Lateglacial period have been removed to the estuaries and river mouths where they have been transported by wave and current action either into deeper water or to contribute to sand and gravel beaches and spits. The valley bottoms have been left in the form of flood plains bounded by one or more terraces. These terraces often provided settlement sites for Bronze Age and Iron Age peoples and still remain the only useful agricultural land in many Highland glens.

The record of sediment transport from both the valley side slopes and the valley bottoms remains one of the topics requiring much more detailed study. It is in the terrace deposits, the many alluvial fans, the occasional lacustrine or marine delta and in the lake bottom deposits that the record of sediment yield (rates of erosion) and environmental conditions in these catchments, will be found. Such work in Scotland is in its infancy (Pennington *et al* 1972, Lovel *et al* 1973, Al-Ansari & McManus 1979, Dickson *et al* 1978, Edwards and Rowntree 1980, Ledger *et al* 1980). Already it has been established that forest clearance and agricultural activity have played an important part in altering erosion and sedimentation rates. In Loch Lomond (Dickson *et al* 1978), massive sediment inputs occurred about 1700 bp and

Plate 6.1. Raised beach—A'Chleit, Kintyre.

during the last 300 years. At Braeroddach Loch in the Dee Valley, Aberdeenshire, Edwards & Rowntree (1980) state that about 4500 bp, as a result of forest clearances by man, there was a threefold increase in the sediment deposition rate from 0·0020 to 0·0072 g cm^{-2} a^{-1}. They also identified greatly accelerated erosion during historical times (p. 220): "About one quarter of the sediment deposited during the last 10,645 calendar years would appear to have been deposited within the last few hundred years." While such studies reveal the significance of human impact on vegetation and rates of erosion and deposition, the separation of natural and man induced changes remains problematical. The conversion of rates of sediment yield to changes in catchment morphology over this sort of timescale also remains difficult to assess.

Apart from the creation of new land areas as a result of the lowering of relative sea-level, changes in the landforms of Scotland over the last 5,000 years have been minimal. A general lowering of the surfaces of individual landforms of one to three metres is not inconceivable particularly if the feature consists of unconsolidated sediments such as till and gravel. Changes in the position of river channels, terrace fronts and coastal dunes and spits of 10–1,000 m are also possible within 5,000 years.

CLIMATE

There is little detailed and reliable evidence of climatic trends in Scotland over the last 5,000 years. Much has been written about the significance of the decline of the forest cover and the expansion of peat and heathlands as an expression of increased wetness and a lowering of temperatures but as will be seen in the discussion of vegetation in the next section, it is difficult to separate the impact of man from any possible natural deterioration in climate as the cause of these changes. It is easy to fall into the trap of circular argument with the botanical evidence currently available for the period 5,000 to 2,000 bp. Many authors claim that climatic changes over the last 10,000 years have been universal in the northern hemisphere and some authors claim that these changes have been synchronous (Lamb 1977, p. 476; Manley 1951; Wendland & Bryson 1974). It has been on the basis of this assumption that the Blytt-Sernander classification (Preboreal, Boreal, Atlantic Subboreal, Subatlantic) and the Godwin-Jessen pollen zones have been regarded as synchronous over wide areas. This leads to the conclusion that synchronous climatic changes are expressed in, and can be recognised in, the biostratigraphy of the Postglacial period. This may be true of the major environmental changes which took place between 13,000 and 5,000 bp in North-west Europe and synchroneity on this continental scale and over this time period may include local variations of plus or minus a thousand years. Now that more accurate dating of environmental changes is possible it seems strange to this writer that chronozones implying the lack of diachroneity of environmental events should be so strongly preferred (Mangerud *et al* 1974). Certainly, in an area the size of the British Isles the expression of what on a global scale may be regarded as a major synchronous climatic change will be translated into changes in vegetation, fauna and rates of erosion and deposition, at different times in different places. Birks (1980) has clearly demonstrated the diachroneity of the major period of expansion of woodland species in Scotland during the period 8,000–4,000 bp. This question of synchroneity of climatic change over wide areas is particularly significant in dealing with the period 5,000 bp to the present.

After the climatic optimum of the Atlantic period (circa 6,000 bp) average temperatures in central England (Fig. 6.1) declined by between 1°C and 2°C so that by about 4,000 bp temperatures were very similar to those recorded for the period 1916–1950 A.D. Since 4,000 bp average winter and summer temperatures in central England have only varied by plus or minus 1°C. There is now a large literature (see bibliographies in Lamb 1972 and Parry 1978) dealing with the interpretation of climatic changes over the last 1,000 years from documentary evidence. Lamb (1972) has produced a table (App. V.3, p. 561) showing winter and summer average temperatures in central England and rainfall in England and Wales (percentage of 1960–50 averages) for each 50 year period from 1100 A.D. to 1950. This table shows that average summer temperatures ranged between 15·8°C and 16·7°C, winter temperatures ranged between 3·1°C and 4·2°C and yearly rainfall ranged between 94 per cent and 106 per cent. Even if it is assumed that these figures are themselves accurate it would be unwise to translate them as being meaningful expressions of temperature and precipitation fluctuations for the whole of Scotland. The range of variation of winter and summer temperatures and of annual precipitation in different regions of Scotland and at different altitudes is sufficiently large to suggest that the minor fluctuations identified for central England need not necessarily be taken as paralleling similar fluctuations in Scotland. A discussion of historical climatic change in Scotland will have to await further studies like that of Parry (1978) in the Lammermuir Hills. Parry does in fact claim that the climatic fluctuations he has identified in the Lammermuir Hills between 1600 and 1800 A.D. do parallel the trends identified in other areas.

Bearing in mind that much of the evidence upon which the following statements are made was obtained from areas outwith Scotland and that the assumption of synchroneity of climatic fluctuations, particularly of small scale changes, has been questioned, it is possible to make some general statements about probable climatic conditions in Scotland over the last 5,000 years.

About 5,000 bp the earliest Neolithic peoples probably enjoyed a climate in which winter temperatures were 1·5°C, and summer temperatures 1°C, warmer than at present and precipitation was some 10–15 per cent more than that of to-day. From 4,000 to 2,000 bp there was a cooling trend with temperatures roughly equivalent to those of to-day but on the basis of botanical evidence, particularly in relation to peat bog development, it has been argued that there was a marked increase in precipitation throughout Scotland. From about 1000 A.D. to 1300 A.D. there was a warming trend (summer temperatures increased by 0·5°C to 0·9°C) and there was a high frequency of warm dry summers and mild winters. From 1300 A.D. until 1700 A.D. there was a deterioration in climatic conditions with increased wetness and storminess with the coldest and wettest summers occurring between 1693 A.D. and 1700 A.D. The period 1800–1900 A.D. saw some improvement but it appears that the climate of Scotland between 1900 A.D. and 1950 A.D. was warmer (by 0·4°C annual average temperature) than at any time since the fifteenth century.

The identification of major climatic changes which would be expressed in concomitant changes in vegetation, soil forming processes and geomorphological processes during the last five thousand years in Scotland is complicated by the unknown extent of the impact of man. Although a world-wide drop in temperatures after 6,000 bp is widely accepted, the direct evidence for such a temperature decline in Scotland is based on the decline in the forest cover which may in part, or even completely, be the product of man-made clearances. The minor fluctuations in average temperatures (plus or minus 1°C) and precipitation (plus or minus ten per cent) have had no significant impact on vegetation, soils and landforms over the last

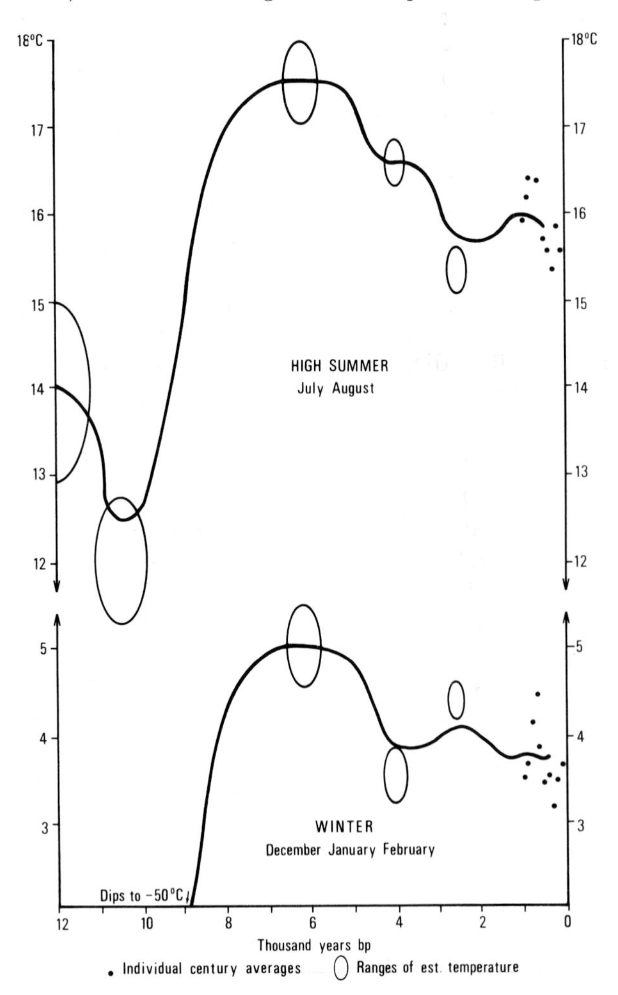

Fig. 6.1.

Average air temperature in lowland Central England. Estimated (or, last millenium, calculated) 1,000 year averages (Lamb 1972).

4,000 bp. This is not to say that a succession of cold winters (e.g. in the seventeenth century A.D.) did not increase the amount of periglacial activity above 800 m and that individual great storms did not extensively damage woodlands or radically change coastal landforms. However, on the basis of presently available evidence the seventeenth century cooling was probably not sufficient for the reestablishment of glaciers in the Scottish Highlands and the thirteenth and fourteenth centuries 'warm' period was not expressed in major vegetational changes. While these minor fluctuations were relatively unimportant in broad environmental terms they may have been highly significant to agricultural practices often being undertaken at the environmental limit for cultivation (Parry 1978).

VEGETATION AND SOILS

There is little doubt that Scotland was a country largely covered by trees about 5000 bp. Although the evidence upon which such a statement is based is the pollen spectra derived from lake sediments and peat bogs and there is some difficulty in translating the pollen evidence into a reconstruction of the character and extent of the forest from which the pollen was derived, general conclusions are possible. These deductions would be more reliable if comparisons of fossil pollen spectra could be made with spectra obtained from similar existing forests in Scotland. This type of information is simply not available in Scotland, largely because of the drastic reduction in the forested area over the last 4,000 years. Studies of modern pollen rain on the European continent (Jonassen 1950; Heim 1962), in forested areas, indicate that in such areas tree pollen ranges between 64 and 92 per cent of the total with a mean of 72 per cent. In most pollen diagrams constructed for Scotland during the maximum extent of the forests, tree pollen values are in the upper part of this range so it can be reliably concluded that the forests were extensive.

Birks, Deacon & Peglar (1975) have constructed isopollen maps for the British Isles for the period immediately before the 'elm decline' (circa 5,000 bp). Figure 6.2 shows the pollen percentages for six species in Scotland about 5,000 bp and it can be deduced that with the exception of the Outer Hebrides, Orkney, Shetland and possibly Caithness, Scotland was heavily forested. These maps show the dominance of pine and birch in the north and north-east, and of oak and hazel in the west and south. Elm percentages of 11–20 only occur in a narrow zone from Ayrshire north-eastwards along the Highland border to Aberdeenshire. The high percentages of alder over much of central Scotland (even though alder is known to be over-represented in pollen diagrams) indicate that large areas of damp lowlands probably with heavy soils, were to be the areas to be occupied by the first agriculturalists to settle in Scotland.

One of the most intriguing and in some ways most difficult questions about the evolution of Scotland's environment is what caused the changes in this extensive forest cover. Major changes did take place in the extent and composition of the forests not only in Scotland (Whittington 1980) but also in England and Wales (Smith 1981) and in continental Europe (Iversen 1944, 1949, 1956, 1969, 1973;

Troels-Smith 1956, 1960). In many pollen diagrams from north-west Europe at the beginning of the Sub-boreal period a decline in several arboreal components, particularly that of elm (Fig. 6.3) and an increase in birch, hazel and some herbaceous taxa (e.g. grasses and ruderals) combined with the appearance of some cereals, all point to major changes taking place in the vegetation cover. These changes have been agglomerated into what has become widely known as the 'elm decline'. Iversen's (1944) work in Denmark pointed to an increased continentality of climate as the originator of these vegetation changes at the Atlantic-Sub-boreal transition and he was able to distinguish later changes as being the product of the effects of Neolithic man on the forest cover. Iversen, therefore, had little difficulty in distinguishing between climatic and anthropogenic effects on the forests. The separation of the Danish Neolithic clearance phases from the elm decline was subsequently challenged by Nilsson (1948) and Troels-Smith (1960). Smith (1981, p. 134) points out that, "In Britain the most marked effects of Neolithic man were subsequently found to coincide with the elm decline."

The pollen diagrams related to the period of the 'elm decline' usually show a decline in tree pollen which is interpreted as an episode of forest clearance. If the associated increase in grass pollen was accompanied by the appearance of cereal pollen then it is suggested that arable agriculture was taking place; when the herbaceous pollen does not include cereals, then only pastoralism was occurring. In some diagrams these changes are succeeded by a return to the former tree pollen percentages and it is deduced that the agricultural activities had been terminated and that regeneration of the forest had occurred.

The nature, extent, dating and significance of these forest clearances has been the subject of much debate amongst archaeologists and palynologists. The openings in the woodland cover were thought to be only of a temporary nature in Iversen's original scheme. His 'landnam' phases were thought to be the product of agricultural activity lasting two or three years which led to soil exhaustion and subsequent abandonment of the site to allow forest regeneration. Troels-Smith (1960) suggested that ivy, elm and mistletoe were used as fodder plants. It has also been suggested that the elm decline was the result of the spread of plant disease.

It is in some ways unfortunate that so much attention has been paid to one species and that the term 'elm decline' has attained so wide a usage. It is now clear (Smith 1970a, p. 91) that tree species other than elm were involved and that changes in the non-arboreal pollen also occurred. The anthropogenic factors cannot be ignored either, and it must be admitted that changes in the 'natural' forests of north-west Europe about 5,000 years ago cannot simply be explained either as the result of climatic change or as the result of human agricultural activities. Smith (1981, p. 127) states: "In considering the Neolithic we shall encounter numerous instances where human activity could have brought about, or contributed to, changes which might otherwise be considered to be climatic effects."

The evidence from sites in Scotland of changes in the nature of the forest cover since 5,000 bp is now quite extensive. Work by H. H. Birks (1972a) and Turner (1964, 1965, 1970) indicate that forest clearance in south-west and central Scotland was relatively late. The absence of radiocarbon dates associated with Birks' work on the Galloway Hills makes an assessment of the chronology of the decline in the tree cover of that area rather difficult but she makes the comment that the evidence from sites in the Galloway Hills is comparable to that of Bloak Moss in Ayrshire (Turner 1970) so that extensive clearances in Galloway probably did not occur until A.D. 400 and it is suggested that complete deforestation did not occur until the eighteenth century. At Bloak Moss, Ayrshire, Turner (1970) has identified small scale clearances, a few hundred square metres in extent and lasting about 50 years, between 3,400 and 3,000 bp. In the Loch Lomond basin (Dickson *et al* 1978),

Fig. 6.2. Isopollen maps of Scotland at 5,000 bp (Birks *et al.* 1975).

substantial changes in the tree cover began about 1,700–1,800 bp. On Flanders Moss in the Forth Valley, Turner (1981) has dated a recurrence surface (2,700 bp) which can be traced right across the bog. The significance of such recurrence surfaces (the junction between two types of peat consisting of different types of vegetation) is currently the subject of considerable debate: it used to be thought that changes in peat stratigraphy clearly indicated changes in climate but the possibility of local drainage, edaphic factors and anthropogenic activity affecting peat formation must be considered. The whole question of the relationship between the decline of the forest cover and the expansion of peat cover and associated moorland development in Scotland either as a direct result of climatic change or of human forest clearance, is still hotly debated. Turner (1981, p. 256) states, "Whilst considerable caution is still required in interpreting the data (i.e. recurrence surfaces and the rate of peat growth), there is no doubt in my mind that the evidence for climatic change in the first millenium BC, from this source, is very strong indeed and cannot be explained simply in terms of local factors." On the other hand Pennington *et al* (1972) and Birks (1972b) have shown that, in the north-west Highlands, moorland and blanket bog had begun to spread at the expense of forest well before the Iron Age.

The evidence of the impact of Neolithic and Bronze Age peoples on the extent of the forest cover and the development of peat/moorland in southern and north-western Scotland remains uncertain. Whereas in the south-west only small scale

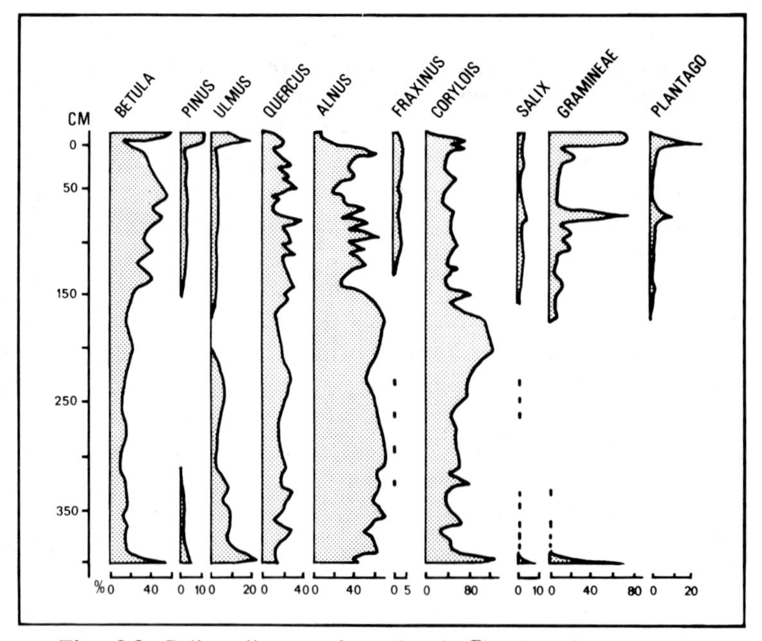

Fig. 6.3. Pollen diagram from South Flanders farm (Turner 1965). Pollen sum is total arboreal pollen excluding *corylus* and *salix*. Neither *calluna* nor *sphagnum* is plotted.

human activity has been established, in the north-west pine and birch woodland was destroyed between 4,000 and 3,500 bp as a result of a combination of climatic deterioration and burning initiated by man (Pennington 1972).

O'Sullivan (1974a, b, 1976) has studied the vegetational history of the Spey Valley and suggests that (1976, p. 301) "The elm decline is not considered as showing any evidence of local human activity in the forest." He also describes the replacement of pine by birch as the main tree type, at circa 3,800 bp, again without any evidence of any marked local disturbance of the forest by man. It was not until 3,400 bp that a forest clearance, lasting about 100 years occurred. It was not until after 3,000 bp that there was a marked influence of man on the vegetation of Speyside, associated with the development of some heathland and the opening of the forest for rough grazing.

The excellent work of Edwards (1975, 1978) and Edwards & Rowntree (1980) at two sites in the Howe of Cromar, in the Dee Valley west of Aberdeen, provides the best available evidence of environmental changes associated with anthropogenic activity in Scotland over the last 5,000 years. Only when many more detailed studies such as this are undertaken will many of the problems raised in this chapter be resolved.

Edwards (1978) obtained sediment cores from two lochs (Braeroddach Loch and Loch Davan) and studied the included pollen and inorganic chemistry, particle size and magnetic susceptibility of the sediments. The two cores contained sediment deposited throughout the Flandrian Age and a total of 34 radiocarbon dates allowed a chronology of sedimentation and vegetational changes to be established. The Howe of Cromar contained a fully-developed forest cover over the period 7,000–5,000 bp consisting of Scots pine with birch, hazel, oak, alder and elm. No evidence of any serious impact by Mesolithic peoples on this forest cover was obtained. By studying changes in the pollen content of the sediments and changes in their chemical constituents and rates of accumulation, Edwards has identified a series of interference and regeneration phases. The first interference phase at Braeroddach Loch has been dated at 5,295 ± 156 bp and that at Loch Davan at 5,105 ± 85 bp. These two dates are inseparable within one standard deviation. The changes recognised in the pollen spectra about 5,200 bp (Fig. 6.4) in this area relate to a drop in all tree pollen, including elm, by 15 per cent of the total pollen. This interference phase lasted 720 radiocarbon years at Braeroddach Loch and 210 radiocarbon years at Loch Davan. Associated with this decline in arboreal pollen, sedimentation rates increased about four-fold—from one centimetre per 50 years to one centimetre per 12 years at Braeroddach Loch. These high sedimentation rates continued for the next 3,000 years (Fig. 6.5). At Braeroddach Loch at least seven interference phases lasting between 100 and 700 years occurred between 5,200 and 2,500 bp and all of them were considerably longer than the typical Danish 'landnam' sequence of 50 years. In an interference phase lasting between 4,300 and 4,100 bp there was a general relative and absolute fall in arboreal pollen with a complementary rise in *Graminae* and other nonarboreal pollen taxa. *Plantago lanceolata* makes its first consistent appearance and *Pteridium aquilinum* has sustained high

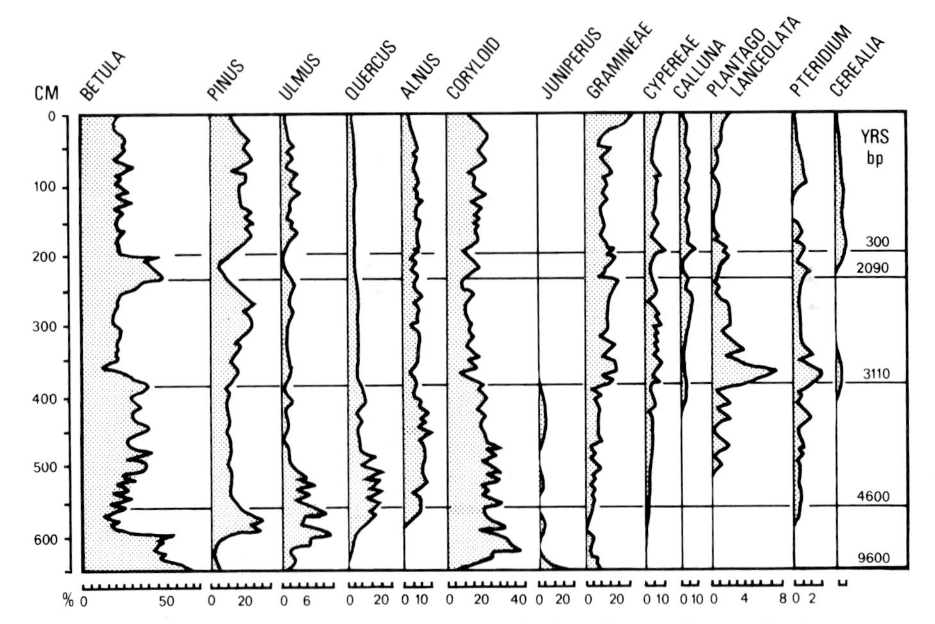

Fig. 6.4. Relative pollen diagram of selected taxa—Braeroddach Loch (Edwards & Rowntree 1980).

values. Edwards is confident that these early interference stages are the product of anthropogenic activity. He states (1978, p. 156), "The high *Plantago lanceolata* values taken together with the grass and bracken frequencies suggest agriculture of a pastoral nature around the site, though arable or mixed activity in the distant parts of the catchment could be contributing to the sustained high sedimentation rate."

The late Neolithic and early Bronze Age (4,500–3,000 bp) can be seen as one of considerable human impact on vegetation and the primarily pastoral activity is linked to the sedimentation rate. During the middle Bronze Age (3,000–2,500 bp) the pollen assemblages reveal a change from tree dominance with the expansion of herbaceous taxa so that by 2,430 bp they are co-dominant. The expansion in *Graminae, Rumex acetosa, Rumex acetosella, Artemisia, Plantago lanceolata* and *Pteridium aquilinum* and many other herbs indicative of arable and pastoral activity along with some cereal pollen (Hardeum type) indicate that from the middle Bronze Age (after 3,000 bp) mixed agriculture with arable cultivation was taking place. This type of activity lasted at least for 2,000 years in this area and may have been continuous, at least to some extent, right through to the present day.

Edwards has clearly established that agricultural activities have affected the vegetation of the Howe of Cromar since about 5,200 bp but such an interpretation does not preclude that changes were also affected as a result of climatic change. From 3,100 bp onwards there was certainly an expansion in the wetland taxa along with *Pinus* (in keeping with its ability to survive on peaty and podsolised soils). Edwards (1978, p. 176) states: "The presence of man during the Bronze and Iron

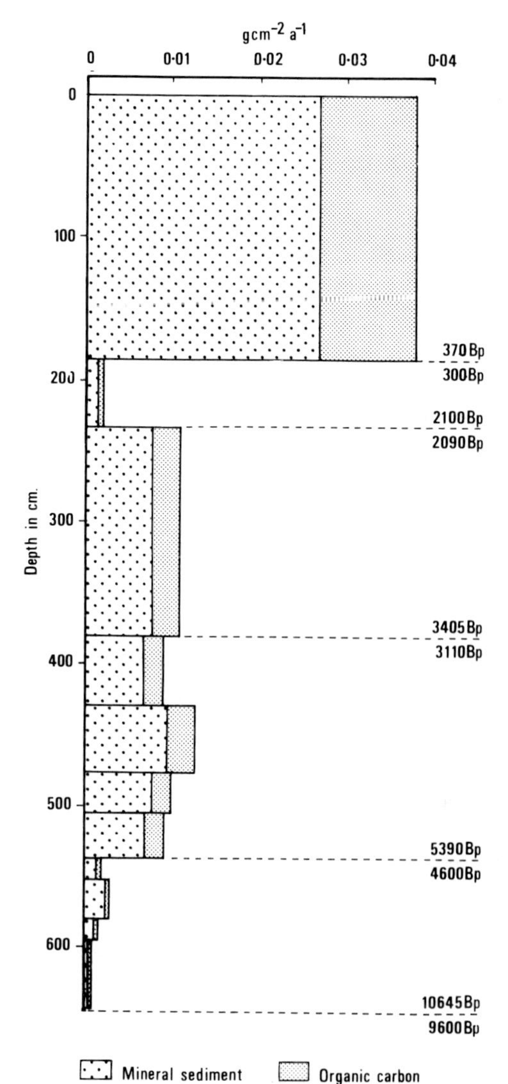

Fig. 6.5.

Rate of sediment deposition Braeroddach Loch (Edwards & Rowntree 1980).

Ages and his ability to bring about edaphic changes, serve to stress the difficulties of demonstrating climatic and/or human causes for the problem. The Cromar sites do not permit any categorical statement in supplying an answer other than to say that during a period of prehistoric activity there occurred a vegetation change compatible with a climatic recession but that the conditions responsible for the expansion of wetland plant types could have stemmed from waterlogging of catchment soils after a long period of human activity and probable soil deterioration."

The transformation of the dominantly tree-covered landscape of Scotland into one in which moorlands dominate is still poorly understood (Fairhurst 1939,

Watson 1939). That this transformation began about 5,000 bp is well established as is the thesis that the demise of the tree cover was accompanied by the spread of peat and increased podsolisation of soils. It is possible that these major changes were initiated by a climatic change leading to increased wetness. The fact that human interference with the vegetation cover has been clearly established, in those areas where both palynological and archaeological investigations have been undertaken, suggests that any natural tendency for a vegetational change would have been stimulated by the clearances made for pastoral or arable activities. Once the tree cover began to decrease the trend to increased podsolisation would be strengthened and even if agricultural activities were terminated the likelihood of forest regeneration taking place would be diminished. The extent of the impact of Neolithic and Bronze Age people on the Scottish landscape remains poorly understood. The problem of the origin of the Scottish moorlands requires many more detailed investigations like that of Edwards in the Howe of Cromar, before it will be resolved.

In some respects, much more is known about vegetation changes during the Neolithic and Bronze Age periods than the ensuing Iron Age, Roman-Iron Age and Pictish periods. Edwards (1978) has suggested that there was a period of woodland regeneration (2,000 bp–A.D. 1100) in the Howe of Cromar but Whittington (1980) states that from about 2,000 bp there was a considerable increase in peat growth and that (p. 34) "Iron Age peoples experienced a climate which was less pleasant than that of the second millenium B.C. and they were operating on a land surface in which peat forming and podsolisation processes were occurring with renewed vigour." He goes on to state (p. 41) that, "By the end of the first millenium A.D. the landscape had undergone radical changes from its Boreal state. Most of the tree cover had at least been modified and in many areas it had been totally removed, being replaced by heath." There is a certain amount of documentary evidence and palynological evidence describing the environment encountered by the Roman invaders. Hanson & Maciness (1980) refer to the date from Bloak Moss, Ayrshire and Flanders Moss in the Forth Valley (Turner 1965), which suggests that those areas were not extensively cleared of forest until 450 A.D. and 200 A.D. respectively. They conclude that forests continued to be a major element in the Scottish landscape at the time of the Roman occupation and on through the period to the fifth century A.D.

This book is primarily concerned with the major natural changes in Scotland's environment. This section has inevitably been concerned with changes in the vegetation cover which may have been initiated by climatic and edaphic factors but which most certainly have been affected by the presence of, and agricultural practices, of man. Over the period 5,000–2,000 bp the size of the human population in Scotland and the nature of the impact of that population's agricultural practices on the vegetation cover remains largely unknown. Most of the studies which have been undertaken reveal that human impact has been very significant within the catchments of the lake basins from which the evidence has been derived.

The general picture revealed by these specific studies will only be clarified by many more detailed investigations both in the fields of palaeoenvironmental reconstruction and archaeology. It is probably safe to state that since 5,000 bp the imprint of man on the evolution of the Scottish vegetation has been more significant than any changes resulting solely from climatic or geological processes. The evolution of the Scottish landscape during the last one thousand years has been almost entirely the product of human action. The deciphering of the story of man's modification of the landscape from the literary and cartographic sources is summarised in a series of essays edited by Parry and Slater (1980) entitled "The Making of the Scottish Countryside."

FAUNA

There was little change in the range of vertebrate species in Scotland in the Neolithic and Bronze Age periods (5,000–3,000 bp) compared with the preceding Mesolithic period. Most of our knowledge of the fauna is derived from archaeological sites and many of these contain bones of red deer (*Cervus elaphus* L) and roe deer (*Capreolus capreolus* L) but no remains of the aurochs (*Bos primigenius* Boj) have been found in Scotland. The wild horse was present in small numbers and the sites on which it has been found have been listed by Grigson (1966). The brown bear (*Ursus arctos*) was widely distributed and walrus bones and teeth have been found at Jarlshof, Shetland and at Northton, Isle of Harris.

The introduction of domesticated animals by the Neolithic peoples led to important changes in the environment. Although cattle and pigs were already present in Britain, Grigson (1966), believed that domesticated cattle and pigs were introduced from Europe and since sheep and goats had no wild forms in Britain they were most certainly imported. Dogs were present in Britain in the Mesolithic period but their significance may have increased with the development of herding practices.

In the previous section on vegetation it was concluded that human activity had modified the forest cover considerably during the period 5,000–2,000 bp. Much of this modification resulted from clearances made for pastoral or arable agriculture. The development of domesticated animal herds would also have affected the vegetation. Cattle are both browsers and grazers and when browsing in woodland they not only damage the trees but trample undergrowth and seedlings. Goats are particularly destructive and certainly assist the process of deforestation. Sheep, however, do require pasture but once it has been created for them their close-cropping of the vegetation discourages any woodland regeneration.

The combination of human clearances and the damage inflicted by domesticated animals certainly led to the diminution of the forests during the Neolithic and Bronze Age periods. However, the presence of the full range of wild woodland mammals during the same period suggests that very large tracts of woodland still survived. By the same token the increasing human population must have put pressure on the wild animal population and the decrease in population of individual species and the number of species present which is characteristic of the last 2,000 years must have begun as early as 5,000 bp.

Numerous species have been eliminated from Scotland during the last 5,000 years. The lemming, the northern lynx and the rat vole disappeared at the dawn of the Christian era and the great elk was eliminated by the destruction of the forests and by hunting (Stephen 1974). The brown bear disappeared in the tenth century but the beaver lasted in the Highlands until the fifteenth and sixteenth centuries while the wild boar survived until the sixteenth century and the wolf until the eighteenth century.

LANDSCAPE CHANGES

The term 'landscape' has been used in a wide variety of contexts with quite a wide range of meanings. In this context it refers to the visual expression of both natural and man-induced changes on the earth's surface. A landscape can be regarded as the momentary expression of many dynamic processes. The major theme of this book has been the identification of the changes over time of the various attributes of the natural environment of Scotland. Over the period 30,000–8,000 bp these changes have been entirely natural. The arrival of Mesolithic hunters and gatherers may have had a small impact on these natural changes but the development of agricultural techniques and a growing human population over the last 5,000 years has made it increasingly difficult to decipher the changes wrought by man as opposed to those produced by natural processes. Little is known about the numbers of people involved in the occupation of Scotland over the period 5,000–1,000 bp. Lythe & Butt (1975) estimate the population of Scotland at various dates as follows:

Late 11th century A.D.	250,000
Late 14th century A.D.	400,000/470,000
Late 16th century A.D.	550,000/800,000
circa 1700 A.D.	800,000/1,000,000

It is difficult to estimate the extent of the human impact in the period up to 1000 A.D. of a population of less than a quarter of a million. The demand for agricultural land certainly led to inroads on the natural forest cover during that period. What is certain is that over the period 1000 A.D. to the present day natural changes in the environment have been far less important in changing the Scottish landscape than the activities of man. Although the initiation of the great moorland areas at the expense of the natural forest cover may or may not have been a purely natural process the presence of a growing population highly dependent on arable and pastoral agriculture led to the creation of the Scottish rural landscape as we now know it. The nineteenth and twentieth centuries saw a further five-fold population increase, much of it in the urban context. Once again a new type of landscape modification was developed in association with the demands of the industrialised society. Control of both the rural and urban environment has now developed to a high degree. The changes produced in the Scottish landscape by man over the last 500 years are far greater than those produced by natural processes. It is only when the rate at which natural processes operate deviates radically from normal that the natural environment reasserts itself and what in human terms is described as a natural disaster occurs. Some of these so called natural disasters are in fact the product of human interference with the natural environment. Landslides, floods, coastal erosion, and subsidence all occur naturally but may, in certain circumstances, be increased both in frequency and magnitude by human interference in the landscape. Five hundred years is a very brief period in the evolution of Scotland's environment and the understanding of man's role in modifying that environment is vital in the cause of sensible planning for the future. In five thousand years the changes in the Scottish environment have been dramatic and largely the product of human interference. Over 30,000 years the environment has been changed on several occasions on a magnitude far greater than that produced by human intervention over the last five thousand years. The magnitude and frequency of environmental changes produced both by natural change and by human impact identified in this book require much more detailed study.

7

THE MAGNITUDE AND FREQUENCY OF ENVIRONMENTAL CHANGES—RETROSPECT AND PROSPECT

Throughout this book, air temperature has been used as an indicator of environmental change because it can be deduced from a wide variety of evidence and at the same time is a good indicator of the general character of a particular range of environmental attributes. Because of the severity of the last glaciation in Scotland there is very little sedimentary and palaeontological evidence of environmental conditions within the country prior to 30,000 bp, but it is possible, by using evidence from other areas, to reconstruct in general terms the broad environmental trends of the last 130,000 years. Most of this book is concerned with the last 30,000 years and that period may be regarded as a model of Scottish environmental changes for about one quarter of a glacial/interglacial sequence. Glacial episodes have been recognised on a global basis (Fig. 7.1A)—eight major periods of ice expansion have been identified during the last 700,000 years and seventeen glacial episodes during the last 1·6 million years. Each glacial episode lasted approximately 100,000 years and each interglacial episode approximately 10,000 years. It therefore seems likely that Scotland's environment throughout the Quaternary Period consisted of at least seventeen major fluctuations, with cold (glacial/periglacial) conditions lasting ten times as long as temperate (interglacial) conditions similar to those of the present. It does not seem unreasonable, therefore, to use the environmental conditions deduced from Scotland during the last 130,000 years as a model for the probable changes which took place at each of the interglacial-glacial-interglacial fluctuations recognised on a global scale.

The record of environmental changes in Scotland during the last 130,000 years can be expressed by estimated mean January and July temperature curves (Fig. 7.1B). The actual temperature values may be in error by ±2°C and the chronology, particularly for the period 130,000–30,000 bp may be in error by several thousand years. However, the onset and termination of the Ipswichian Interglacial (125,000–115,000 bp) seem to have been as rapid as the onset of the present interglacial (10,000 bp to present). Little direct evidence is available about the vegetation, soils, fauna and sea-levels of the Ipswichian Interglacial in Scotland, but they are likely to have been broadly similar to those of the present interglacial prior to human modification. Since human occupation occurred in England during the Ipswichian Interglacial it might be expected that Scotland was also occupied by man at that time, but as yet no archaeological evidence for this period has been found in Scotland.

The Devensian glacial episode in Scotland prior to 30,000 bp is only poorly known and little more can be said than that it was probably characterised by long periods of cold dry winters (mean January temperatures of −5° to 0°C) and cool dry summers (mean July temperatures 5° to 10°C) interspersed with short intervals (interstadials) of about 3,000 years when mean monthly temperatures were 3–5°C warmer. Two major questions require to be answered about this period of time in

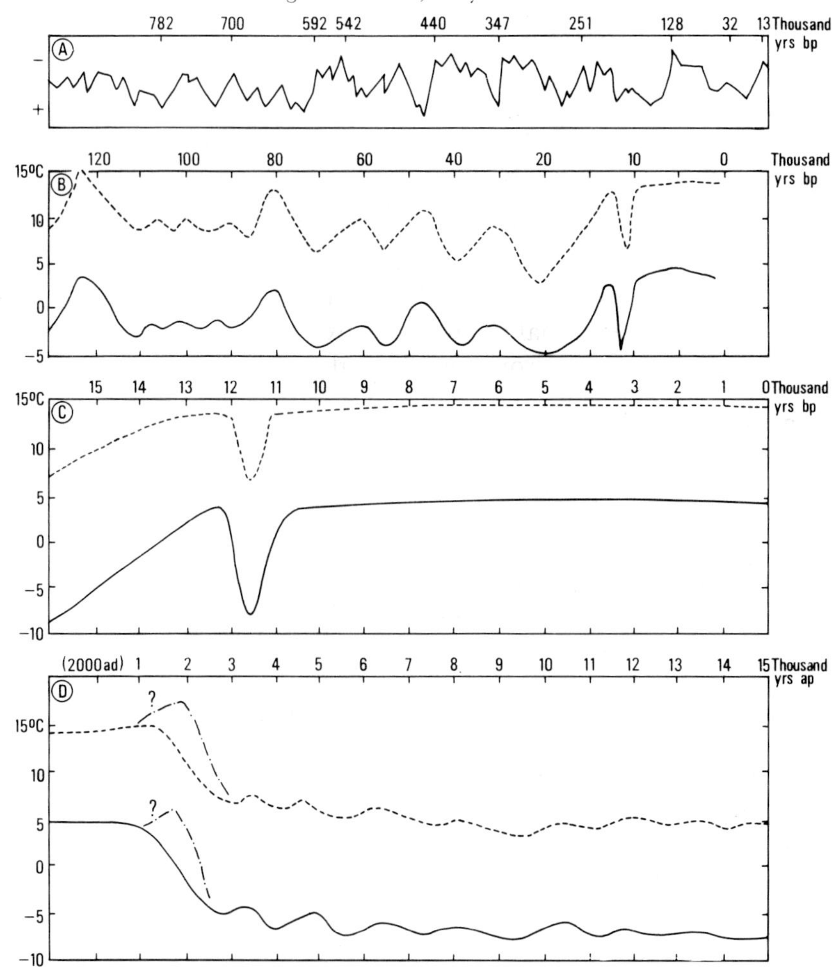

Fig. 7.1. A. Estimated global ice volumes. (Shakleton & Opdyke 1973).
B. Estimated sea-level air temperatures, January and July, in Scotland during the last 130,000 years.
C. Estimated sea-level air temperatures, January and July, in Scotland during the last 15,000 years.
D. Estimated sea-level air temperatures, January and July, in Scotland during the next 15,000 years.

Scotland. Firstly, were the colder episodes conducive to the build up and expansion of glaciers and, secondly, were the interstadials sufficiently warm and long enough to allow the development of a forest cover? Associated with an answer to the first question will be a greater understanding of relative sea-level fluctuations between 130,000 and 30,000 bp because, if there was no significant glacier development in Scotland during this period then as the global ice volume increased, relative sea-level would be continually falling around the Scottish coast and by the time the late Devensian ice sheet was established (after 30,000 bp), sea-level may have been 100 m lower than at present. If, on the other hand, glacier complexes did develop in Scotland on one or more occasions during the early and middle Devensian, then isostatic depression and uplift would have produced several fluctuations in relative sea-level.

The chronology of the expansion of the last ice sheet (Late Devensian) to cover Scotland is still a matter of debate. The effects of that ice sheet and the environmental changes which accompanied its wastage are better known. Between 15,000 and 13,000 bp (Fig. 7.1C) air temperatures began to rise rapidly (by 4°C in less than 1,000 years). This was a false beginning to the present interglacial. Sometime after 12,000 bp temperatures began to rise more slowly or to decline (depending on the interpretation of different lines of evidence) but after 11,000 bp there was a rapid decline in mean monthly temperature (6–10°C in less than 500 years) before an equally rapid rise to initiate the present interglacial about 10,000 bp. The Lateglacial period in Scotland (14,000–10,000 bp) was characterised by rapid and very significant environmental changes. The land areas, as they became ice free, were soon colonised by plants and animals and relative sea-level, which was initially high in central Scotland (+40 m), was continuously falling throughout the period. The Lateglacial Interstadial (14,000–11,000 bp) was not of sufficient duration for the heath-grassland-dwarf shrub association to be replaced by a forest cover. The general amelioration of the climate ceased abruptly about 11,000 bp with the onset of the Loch Lomond Stadial associated with a return to glacial and periglacial conditions. The recognition of this event is not only based on sedimentary and palaeontological evidence from Scotland but from other parts of north-west Europe and the north-east Atlantic ocean. There can be little doubt that the magnitude of the environmental changes which took place during the Loch Lomond Stadial was considerable, but the chronology of those changes is difficult to establish with the presently-available radiometric dating techniques. It would appear that changes in mean January temperature of the order of 10°C took place in less than 1,000 years and may be in less than 500 years. It has been demonstrated that such dramatic climatic changes in Scotland are related to major changes in the position of the 'polar front' separating the relatively cold from the relatively warm waters in the north-east Atlantic Ocean.

The present interglacial began approximately 10,000 bp in Scotland and within 3,000 years the country was covered by a birch-oak-pine forest. There was a significant diachronous marine transgression which culminated between 6,000 and 5,000 bp and which affected large areas of central lowland Scotland up to altitudes of 15 m above present sea-level. Throughout the present interglacial (Holocene/Flandrian) mean air temperatures at sea-level have not fluctuated by more than 2°C and precipitation has probably not varied by 15 per cent. The magnitude of these variations is insignificant in terms of the general trends examined in this book and they are less significant than the impact on the natural environment of the increasing human population of Scotland after 5,000 bp.

The magnitude and frequency of environmental changes in Scotland during the last 30,000 years may be summarised as follows. After a period of some 15,000 years (30,000–15,000 bp) during which air temperatures at sea-level were at least 10°C cooler than at present and a major ice sheet covered Scotland, climatic amelioration was rapid and the ice sheet wasted away quickly. Relative sea-levels dropped in some areas by 40 m in about 3,000 years. The Lateglacial (Loch Lomond) stadial was characterised by climatic changes of large magnitude which operated over a very short period of time (less than 1,000 years). The present interglacial was initiated rapidly and has been relatively stable apart from relative sea-level changes of less than 15 m in 6,000 years and the woodland clearances carried out by man.

Based largely on evidence from England and the eastern part of the North Atlantic Ocean it can be suggested that environmental changes similar to those which took place between 13,000 and 10,000 bp, but not necessarily of the same magnitude, took place on at least four occasions between 30,000 and 120,000 bp (30,000 bp; 48,000 bp; 60,000 bp; 80,000 bp). It would seem therefore, that it is much too simplistic to consider a 'glacial period' as having a fairly uniform environment. It would appear that the magnitude and frequency of environmental changes in an interglacial episode are far less than those that occur within a glacial episode. However, the magnitude of environmental changes based on evidence of mean monthly air temperature fluctuations at sea-level involve the consideration of thresholds. Specifically, the operation of periglacial processes, the establishment and expansion of glaciers, and the colonisation of an area by certain plants and animals, all require the crossing of certain temperature and precipitation thresholds and the maintenance of these conditions for a certain period of time. The recognition of these threshold crossings in the sedimentary and palaeontological record is only in its infancy in Scotland. Similarly, the recognition of regional variations within Scotland of palaeoenvironments is hardly justified, except in very general terms, on the basis of the presently available evidence. Since altitudinal variations and 'west-east' contrasts are distinguishable in the present interglacial environment, it is usually assumed that similar internal variations existed during glacial episodes. However, it might be argued that the lowering of snow-lines and an increase in the 'continentality' of Scotland's climate might eradicate some of the regional and altitudinal contrasts that now exist.

PROSPECT FOR THE FUTURE

Largely as a result of the analysis of deep ocean sediments it has been concluded that the climatic changes associated with the sequence of glacial and interglacial episodes during the Quaternary period are global in extent and are largely controlled by changes in the Earth's eccentricity, precession and tilt. However, although the occurrence of glacial episodes approximately every 100,000 years has been established and all the evidence points to interglacial episodes only lasting approximately 10,000 years, factors other than those determined by astronomical relationships can be important in determining the timing of local terminations of glacial or interglacial episodes. The importance of feed-back mechanisms such as the release of great quantities of meltwater from melting ice sheets into a warming ocean or the impact of sea-ice or an ice sheet on precipitation sources, must also be considered. It is therefore with some caution that the limited data on palaeoenvironments should be used to discuss possible future trends.

On the basis of a 100,000 year cycle for glacial episodes separated by 10,000 year long interglacial episodes, the present interglacial is due to end within the next few thousand years. However, it has been argued that because of the great impact of the burning of fossil fuels and the associated production of carbon dioxide there may well be a worldwide rise in air temperature during the next 1,000 years to produce a 'super-interglacial'. In other words the present interglacial may last longer than would be expected simply on the basis of the Quaternary record. The impact of increased carbon dioxide content in the atmosphere on the astronomically induced climatic changes is still a matter of debate.

The climatic changes deduced for Scotland during the last 130,000 years have been closely linked with the position of the polar front in the North Atlantic Ocean. It is known that over the period 11,000–10,000 bp the front migrated from western Iceland to the Bay of Biscay producing changes in mean monthly air temperatures at sea-level in Scotland of between 6° and 10°C. If a similar change in the oceanic circulation takes place in the near future then a temperature change of the order of 1·5°C per century might be expected and within 200 years glaciers will develop on the high ground of Scotland. A general cooling of average global temperature has been recognised since 7,000 bp and the Quaternary record would suggest that it is highly likely that sometime within the next few thousand years the present interglacial environment in Scotland will terminate rapidly. Permanent snow beds will expand into glaciers and within 3,000 years much of Scotland will be covered by an ice sheet. This process could be postponed by the higher content of carbon dioxide in the atmosphere or by the slower southward migration of the ocean polar front. At the beginning of the last major glacial episode (Early Devensian) there probably was a 30,000 year period of cold dry conditions before any glaciers developed in Scotland. Based on the magnitude and chronology of temperature changes which took place between 11,000 and 10,000 bp however, the return of glacial conditions to Scotland may well be a dramatic event.

FUTURE RESEARCH

The existing literature on Scottish Quaternary palaeoenvironment is largely the product of some excellent specialist studies in a wide variety of fields. This book has attempted to place each of these specialist contributions into a framework of environmental changes so that the magnitude and frequency of those changes can be assessed. Much more basic data are still required, but only when these data can be placed into an accurate absolute chronology will they become meaningful in terms of an understanding of the rate of change operating in both the natural and man-modified environments. In terms of the last 30,000 years, radiocarbon dating has been heavily relied upon and it appears that it has major weaknesses as a technique for dating short term changes. New and better absolute dating techniques are urgently required. Similarly, the impact of man in modifying the natural environment and inducing environmental changes requires much greater attention. The interdisciplinary approach, which is so characteristic of all aspects of Quaternary studies, needs to be strengthened if the development in our understanding of Scotland's environment, achieved during the last twenty years, is to be deepened.

BIBLIOGRAPHY

Adhemar, J. A. 1842. *Revolutions de la mer*. Paris.

Agassiz, L. 1840. On the evidence of the former existence of glaciers in Scotland, Ireland and England. *Proc. Geol. Soc.*, Lond. **3,** 327–32.

Al-Ansari, N. A. & McManus, J. 1979. Fluvial sediments entering the Tay Estuary: sediment discharge from the River Earn. *Scott. J. Geol.* **15,** 203–16.

Anderson, J. G. C. 1949. The Gare Loch readvance moraine. *Geol. Mag.* **86,** 239–44.

Andrews, J. T. & Barry, R. G. 1978. Glacial inception and disintegration during the last glaciation. *Ann. Rev. Earth. Planet. Sci.* **6,** 205–28.

Atkinson, R. J. C. 1962. Fisherman and Farmers. In Piggott, S. (ed.) *The prehistoric peoples of Scotland*, 1–38. London.

Bailey, E. B. *et al* 1924. Tertiary and Post-Tertiary geology of Mull, Loch Aline and Oban. *Mem. Geol. Surv. Scot.*

Bald, M. 1809. On the old alluvial cover of Clackmannan. *Wern. Mems.* **1,** 841.

Ballantyne, C. K. 1979. A sequence of Lateglacial ice-dammed lakes in East Argyll. *Scott. J. Geol.* **15,** 153–60.

Ballantyne, C. K. & Waine-Hobson, T. 1980. The Loch Lomond Advance on the Island of Rhum. *Scott. J. Geol.* **16,** 1–10.

Barrett, J. *et al* 1976. Second millenium B.C. banks in the Black Moss of Achnacree. *Brit. Archaeol. Rep.* **33,** 283–7.

Bartley, D. D. 1962. The stratigraphy and pollen analysis of lake deposits near Tadcaster, Yorkshire. *New Phytol.* **61,** 277–87.

Bell, D. 1877. On the aspects of Clydesdale during the glacial period. *Trans. Geol. Soc. Glasgow.* **4,** 66.

Bennie, J. 1896. On the occurrence of peat with arctic plants in boulder clay at Faskine, near Airdrie. *Trans. Geol. Soc. Glasgow.* **10,** 148–52.

Berger, A. 1977. Support for the astronomical theory of climatic change. *Nature,* Lond. **269,** 44–45.

Berger, A. L. 1978. Long term variations of caloric insolation resulting from the earth's orbital elements. *Quat. Res.* **9,** 139–167.

Berglund, B. E. 1979. The deglaciation of southern Sweden 13,500–10,000 BP. *Boreas,* **8,** 89–118.

Binns, P. E. *et al* 1973. The geology of the Sea of the Hebrides. *Rep. Inst. Geol. Sci.* No. 73/14.

1974. Glacial and post-glacial sedimentation in the Sea of the Hebrides. *Nature*, Lond. **248,** 751–4.

Birks, H. H. 1970. Studies in the vegetational history of Scotland I. A pollen diagram from Abernethy Forest, Inverness-shire. *Journ. Ecol.* **58,** 827–46.

1972a. Studies in the vegetational history of Scotland II. Two pollen diagrams from the Galloway Hills, Kircudbrightshire. *Journ. Ecol.* **60,** 183–217.

1972b. Studies in the vegetational history of Scotland III. A radiocarbon-dated pollen diagram from Loch Maree, Ross and Cromarty. *New Phytol.* **71,** 731–54.

1973. Modern macrofossil assemblages in lake sediments in Minnesota. In Birks, H. J. B. & West, R. G. (eds.), *Quaternary plant ecology*, Oxford, 1973–89.

Birks, H. H. & Birks, H. J. B. 1980. *Quaternary palaeoecology*. London.

Birks, H. H. & Matthewes, R. W. 1978. Studies in the vegetational history of Scotland V. Late Devensian and early Flandrian pollen and macrofossil stratigraphy at Abernethy Forest, Inverness-shire. *New Phytol.* **80,** 455–84.

Birks, H. J. B. 1973a. Modern pollen rain studies in some arctic and alpine environments. In Birks, H. J. B. & West, R. G. (eds.), *Quaternary plant ecology*. Oxford, 143–68.

1973b. *Past and present vegetation of the Isle of Skye: a palaeoecological study*. Cambridge.

1977. The Flandrian forest history of Scotland: a preliminary synthesis. In Shotton, F. W. (ed). *British Quaternary studies: recent advances*. Oxford.

1980. Quaternary vegetational history of west Scotland. *International palynological conference excursion guide.* Cambridge.

BIRKS, H. J. B. & RANSOM, M. E. 1969. An interglacial peat at Fulga Ness, Shetland. *New Phytol.* **68,** 777–96.

BIRKS, H. J. B. *et al* 1975. Pollen maps for the British Isles 5000 years ago. *Proc. Roy. Soc. Lond.* (B) **189,** 87–105.

BISHOP, W. W. 1963. Late-glacial deposits near Lockerbie, Dumfriesshire. *Trans. J. Proc. Dumfries. Galloway, Nat. Hist. Antiq. Soc.* 4D, 117–32.

BISHOP, W. W. & COOPE, G. R. 1977. Stratigraphical and faunal evidence for Late-glacial and early Flandrian environments in south-west Scotland. In Gray, J. M. & Lowe, J. J. (eds.) *Studies in the Scottish Lateglacial environment,* 61–88, Oxford.

BISHOP, W. W. & DICKSON, J. H. 1970. Radiocarbon dates related to the Scottish Late-glacial sea in the Firth of Clyde. *Nature* **227,** 480–82.

BLACKBURN, K. B. 1946. On a peat from the island of Barra, Outer Hebrides. Data for the study of post-glacial history. *New Phytol.* **45,** 44–49.

BLYTT, A. 1876. *Essay on the immigration of the Norwegian flora during alternating rainy and dry periods.* Christiana.

BOULTON, G. S. *et al* 1977. A British ice sheet model and patterns of glacial erosion and deposition in Britain. In Shotton, F. W. (ed.) *British Quaternary studies: recent advances.* Oxford.

BOWEN, D. Q. 1977. Hot and cold climates in prehistoric Britain. *Geog. Mag.* **69,** 685–98.

1978. *Quaternary geology.* Oxford.

1979. Geographical perspective on the Quaternary. *Prog. in Phys. Geog.* **3,** 167–86.

1980. The Quaternary of the United Kingdom. in Owen, T. R. (ed.) *United Kingdom: introduction to general geology and guides to excursions,* 88–91, I.G.C. Paris.

BRADY, G. S., CROSSKEY, H. W. & ROBERTSON, D. 1874. The post-Tertiary Entomostraca of Scotland. *Mon. Pal. Soc. Lond.,* 229 pp.

BREMNER, A. 1932. Further problems in the glacial geology of north-eastern Scotland. *Trans. Edinb. Geol. Soc.* **12,** 147–64.

1934a. The glaciation of Moray and ice movements in the north of Scotland. *Trans. Edinb. Geol. Soc.* **13,** 17–56.

1934b. Meltwater drainage channels and other glacial phenomena of the Highland Border Belt from Cortachy to the Bervie Water. *Trans. Edinb. Geol. Soc.* **13,** 174–75.

1939. The Late-glacial geology of the Tay Basin from Pass of Birnam to Grantully and Pitlochry. *Trans. Edinb. Geol. Soc.* **13,** 473–74.

BRETT, D. W. & NORTON, T. A. 1969. Late-glacial marine algae from Greenock and Renfrew. *Scott. J. Geol.* **5,** 42–48.

BROECKER, W. S. & VAN DONK, J. 1970. Insolation changes, ice volumes and the ^{18}O record in deep-sea cores. *Rev. Geophys. and Space Phys.* **8,** 169–98.

BROOK, C. L. 1972. Pollen analysis and the Main Buried Beach in the western part of the Forth Valley. *Trans. Inst. Br. Geogr.* **55,** 161–70.

BROWN, M. A. E. *et al* 1977. The date of the deglaciation of the Paisley-Renfrew area. *Scott. J. Geol.* **13,** 301–3.

BROWN, M. A. E. 1980a. In Jardine, G. W. (ed.) *Quat. Res. Assoc. Field Guide—Glasgow Region,* 11–13, Glasgow.

1980b. Late Devensian marine limits and the pattern of deglaciation of the Strathearn area, Tayside. *Scott. J. Geol.* **16,** 221–30.

BROWN, M. A. E. & GRAHAM, D. K. 1981. Glaciomarine deposits of the Loch Lomond stade glacier in the Vale of Leven between Dumbarton and Balloch, west central Scotland. *Quat. Newsletter* **34,** 1–7.

BROWN, M. A. E. *et al* 1981. New evidence for late Devensian marine limits in east-central Scotland. *Quat. Newsletter* **34,** 8–15.

BRYCE, J. 1873. Footnote. *Q. J. Geol. Soc. Lond.* **29,** 545.

BUCKLAND, W. 1840. On the former existence of glaciers in Scotland. *Edinb. New Phil. J.* **30,** 194–98.

BURKE, M. J. 1969. The Forth Valley: an ice-moulded lowland. *Trans. Inst. Br. Geogr.* **48,** 51–9.

BURNETT, J. H. (ed.) 1964. *The vegetation of Scotland.* Edinburgh.

CAMPBELL, J. F. 1873. Notes on the glacial phenomena of the Hebrides. *Q. J. Geol. Soc. Lond.* **29,** 545–48.

CASELDINE, C. J. 1979. Early land clearance in south-east Perthshire. In Thoms, L. M. (ed.) *Early man in the Scottish landscape.* Scottish Archaeological Forum 9, 1–15. Edinburgh.

1980. A Lateglacial site at Stormont Loch, near Blairgowrie, eastern Scotland. In Lowe, Gray & Robinson (eds.) *Studies in the Lateglacial of north-west Europe*, 69–88. Oxford.

Caseldine, C. J. & Edwards, K. J. 1982. Interstadial and last interglacial deposits covered by till in Scotland: comments and new evidence. *Boreas* **11,** 119–22.

Chamberlin, T. C. 1895. Glacial phenomena of North America. In Geikie, J. *The great ice age*, 724–55. New York.

Chambers, R. 1848. *Ancient sea margins*. Edinburgh.

Charlesworth, J. K. 1926a. The glacial geology of the Southern Uplands, west of Annandale and upper Clydesdale. *Trans. R. Soc. Edinb.* **55,** 1–23.

1926b. The readvance marginal kame-moraine of the south of Scotland and some later stages of retreat. *Trans. R. Soc. Edinb.* 25–50.

1955. Lateglacial history of the Highlands and Islands of Scotland. *Trans. R. Soc. Edinb.* **62,** 769–928.

Chester, D. K. 1980. The evaluation of Scottish sand and gravel resources. *Scott. Geogr. Mag.* **96**(1), 51–62.

Clapperton, C. M. 1968. Channels formed by the superimposition of glacial meltwater streams, with special reference to the East Cheviot Hills, northeast England. *Geogr. Annlr.* **50,** 207–20.

1971. The location and origin of glacial meltwater phenomena in the eastern Cheviot Hills. *Proc. York. Geol. Soc.* **38,** 361–80.

Clapperton, C. M. *et al* 1975. Loch Lomond Readvance in the eastern Cairngorms. *Nature* **253,** 710–12.

Clapperton, C. M. & Sugden, D. E. 1977. The Aberdeen and Dinnet glacial limits reconsidered. In Clapperton, C. M. (ed.) *North-east Scotland: geographical essays,* 5–11. Abderdeen.

Clark, J. G. D. 1956. Notes on the Obanian with special reference to antler and bone-work. *Proc. Soc. Antiq. Scot.* **89,** 91–106.

Cline, R. M. & Hays, J. D. 1976. Investigations of Late Quaternary palaeo-oceanography and palaeoclimatology. *Geol. Soc. Am. Mem.* **145,** 77–110.

Clough, C. T. *et al* 1910. The geology of East Lothian. *H.M. Geol. Surv. London.* Mem. 2nd Ed.

Coles, J. M. 1971. The early settlement of Scotland: excavations at Morton, Fife. *Proc. Prehist. Soc.* **37,** 284–366.

Connell, R. *et al* 1982. Evidence for two pre-Flandrian palaeosols in Buchan, north-east Scotland. *Nature* **297,** 570–72.

Coope, G. R. 1968. Fossil beetles collected by James Bennie from Lateglacial silts at Corstorphine, Edinburgh. *Scott. J. Geol.* **4,** 339–48.

1975. Climatic fluctuations in north-west Europe since the last Interglacial indicated by fossil assemblages of coleoptera. In Wright, A. E. & Moseley, F. (eds.) *Ice ages ancient and modern,* 153–68. Geol. J. Spec. Issue 6, Liverpool.

1977. Fossil coleopteran assemblages as sensitive indicators of climatic changes during the Devensian (last) cold stage. *Phil. Trans. R. Soc. Lond.* B. **280,** 313–337.

Coope, G. R. *et al* 1961. A late Pleistocene fauna and flora from Upton Warren, Worcestershire. *Phil. Trans. R. Soc. Lond.* B. **244,** 379–421.

Coope, G. R. & Pennington, W. 1977. The Windermere Interstadial of the Late Devensian. *Phil. Trans. R. Soc. Lond.* B. **280,** 337–39.

Cormack, W. F. 1970. A mesolithic site at Barsalloch, Wigtownshire. *Trans. Dumfries. Galloway Nat. Hist. Antiq. Soc.* **47,** 63–80.

Cormack, W. F. & Coles, J. M. 1968. Mesolithic site at Low Clone, Wigtownshire. *Trans. Dumfries. Galloway Nat. Hist. Antiq. Soc.* **45,** 44–72.

Cornish, R. 1981. Glaciers of the Loch Lomond Stadial in the western Southern Uplands of Scotland. *Proc. Geol. Ass.* **92,** 105–14.

Coward, M. P. 1977. Anomolous glacier erratics in the southern part of the Outer Hebrides. *Scott. J. Geol.* **13,** 185–88.

Cox, A. *et al* 1963. Geomagnetic polarity epochs and Pleistocene geochronometry. *Nature,* Lond. **198,** 1049–51.

1964. Reversals of the earth's magnetic field. *Science* **144,** 1537–43.

Craig, G. Y. (ed.) 1970. *The geology of Scotland.* Edinburgh.

Croll, J. 1875. *Climate and Time.* New York.

Cullingford, R. A. 1972. Lateglacial and Postglacial shoreline displacement in the Earn-Tay area and eastern Fife. *Univ. of Edinburgh Ph.D. thesis* (unpubl.).

1977. Lateglacial raised shorelines and deglaciation in the Earn-Tay area. In Gray, J. M. & Lowe, J. J. (eds.) *Studies in the Scottish Lateglacial environment*, 15–32. Oxford.

CULLINGFORD, R. A. & SMITH, D. E. 1966. Lateglacial shorelines in eastern Fife. *Trans. Inst. Br. Geogr.* **39,** 31–51.

1980. Late Devensian shorelines in Angus and Kincardineshire, Scotland. *Boreas* **9,** 721–38.

CULLINGFORD, R. A. *et al* 1980. Early Flandrian sea level changes in lower Strathearn. *Nature* **284,** 159–61.

CUNNINGHAM-CRAIG, E. H. *et al* 1911. The geology of Colonsay and Oronsay, with part of the Ross of Mull. *Mem. Geol. Surv. U.K.*

DARLING, F. F. & BOYD, J. M. 1969. *The Highlands and Islands*. Edinburgh.

DARWIN, C. 1839. Observations on the parallel roads of Glen Roy and of other parts of Lochaber. *Phil. Trans.* 1839, 39.

DAVIDSON, D. A. *et al* 1976. Palaeoenvironmental reconstruction and evaluation: a case study from Orkney. *Trans. Inst. Br. Geogr.* (New series) **1,** 346–61.

DAVIES, G. L. 1968. The tour of the British Isles made by Louis Agassiz in 1840. *Annals of Science* **24,** 131–46.

DAWSON, A. G. 1977. A fossil lobate rock glacier in Jura. *Scott. J. Geol.* **13,** 37–42.

1979. A Devensian medial moraine in Jura. *Scott. J. Geol.* **15,** 43–48.

1980a. The low rock platform in western Scotland. *Proc. Geol. Ass.* **91,** 339–44.

1980b. Raised shorelines and Weichselian ice-sheet decay in the southern Scottish Inner Hebrides. *Quat. Newsletter* 30.

1980c. Shore erosion by frost: an example from the Scottish Lateglacial. In Lowe, J. J., Gray, J. M. & Robinson, J. E. (eds.) *Studies in the Lateglacial of North-west Europe*, 45–54. Oxford.

DEEGAN, C. E. *et al* 1973. The superficial deposits of the Firth of Clyde and its sea lochs. *Rep. Inst. Geol. Sci.* 73/9, 42 pp.

DENTON, G. H. & HUGHES, T. J. (eds.) 1981. *The last great ice sheets*. New York.

DENTON, G. H. & KARLEN, W. 1973. Holocene climatic variations—their pattern and possible cause. *Quat. Res.* **3,** 155–205.

DICKSON, J. H. *et al* 1976. Three Late-Devensian sites in west central Scotland. *Nature*, Lond. **262,** 43–44.

1978. Palynology, palaeomagnetism and radiometric dating of Flandrian marine and freshwater sediments of Loch Lomond. *Nature*, Lond. **274,** 548–53.

DIMBLEBY, G. W. 1961. The ancient forest of Blackmore. *Antiquity* **35,** 123–8.

1962. The development of British heathlands and their soils. *Oxford Forestry Mem.* 23.

DOBSON, M. R. & EVANS, D. 1975. Glacial and post-glacial sedimentation in the Malin Sea. *Geol. Soc. Lond.* Unpubl. contribution.

DONNER, J. J. 1957. The geology and vegetation of Lateglacial retreat stages in Scotland. *Trans. R. Soc. Edinb.* **63,** 221–64.

1970. Land/sea level changes in Scotland. In Walker, D. & West, R. G. (eds.) *Studies in the vegetational history of the British Isles*, 23–39. Cambridge.

1979. The early or middle Devensian peat at Burn of Benholm, Kincardineshire. *Scott. J. Geol.* **15,** 247–50.

DUPLESSEY, J. C. *et al* 1981. Deglacial warming of the north-eastern Atlantic Ocean: correlation with the palaeoclimatic evolution of the European continent. *Palaeogeog. Palaeoclim. Palaeoecol.* **35,** 121–44.

DURNO, S. E. 1956. Pollen analysis of peat deposits in Scotland. *Scott. Geogr. Mag.* **72,** 177–87.

1957. Certain aspects of vegetational history in north-east Scotland. *Scott. Geogr. Mag.* **73,** 176–84.

1958. Pollen analysis of peat deposits in eastern Sutherland and Caithness. *Scott. Geogr. Mag.* **74,** 127–35.

1959. Pollen analysis of peat deposits in the eastern Grampians. *Scott. Geogr. Mag.* **75,** 102–11.

EDEN, R. A. *et al* 1978. Quaternary deposits of the central North Sea 6. Depositional environments of offshore Quaternary deposits of the continental shelf around Scotland. *Rep. Inst. Geol. Sci. Lond.* 77/15, 18 pp.

EDWARDS, K. J. 1974. A half-century of pollen analytical research in Scotland. *Trans. Bot. Soc. Edinb.* **42,** 211–22.

1975. Aspects of the prehistoric archaeology of the Howe of Cromar. In Gemmell, A.

M. D. (ed.) *Quaternary studies in north-east Scotland,* 82–87. Aberdeen.

1978. Palaeoenvironmental and archaeological investigations in the Howe of Cromar, Grampian Region, Scotland. *Univ. of Aberdeen Ph.D. thesis* (unpubl.).

EDWARDS, K. J. *et al* 1976. Possible Interstadial and Interglacial pollen floras from Teindland, Scotland. *Nature* **264,** 742–44.

EDWARDS, K. J. & CONNELL, E. R. 1981. Interglacial and Interstadial sites in north-east Scotland. *Quad. Newsletter* **33,** 22–28.

EDWARDS, K. J. & ROWNTREE, K. M. 1980. Radiocarbon and palaeoenvironmental evidence for changing rates of erosion at a Flandrian stage site in Scotland. In Cullingford, R. A. *et al* (eds.) *Timescales in geomorphology,* 207–24. Chichester.

EMILIANI, C. & SHACKLETON, N. J. 1974. The Brunhes Epoch: palaeotemperatures and geochronology. *Science* **183,** 511–14.

ERDTMAN, G. 1923. Jakttagelser fram en mikrpaleontologisk undersokning at Nordskotska, Hebridiska, och Shetlandandska torvmarker. *Geol. Foren. Stockh. Forh.* **45,** 538–45.

1924. Studies in micropalaeontology of post-glacial deposits in northern Scotland and the Scotch Isles, with special reference to the history of the woodlands. *J. Linn. Soc.* **46,** 449–504.

1928. Studies in the Postarctic history of the forests of northwestern Europe. I. Investigations in the British Isles. *Geol. Foren. Stockh. Forh.,* 123–92.

FAIRHURST, H. 1939. The natural vegetation of Scotland: its character and development. *Scott. Geogr. Mag.* **55,** 193–211.

FITZPATRICK, E. A. 1963. Deeply weathered rock in Scotland, its occurence, age and contribution to soils. *J. Soil Sci.* **14,** 33–43.

1965. An Interglacial soil at Teindland, Morayshire. *Nature* **207,** 621–2.

1972. The principal Tertiary and Pleistocene events in north-east Scotland. In Clapperton, C. M. (ed.) *North-east Scotland: geographical essays,* 1–4. Aberdeen.

FLENLEY, J. R. & PEARSON, M. C. 1967. Pollen analysis of a peat from the island of Canna (Inner Hebrides). *New Phytol* **66,** 299–306.

FLINN, D. 1978a. The glaciation of the Outer Hebrides. *Geol. J.* **13,** 195–99.

1978b. The most recent glaciation of the Orkney-Shetland Channel and adjacent regions. *Scott. J. Geol.* **14,** 109–123.

1978c. The erosional history of Shetland: a review. *Proc. Geol. Ass.* **88,** 129–46.

FLINT, R. F. 1971. *Glacial and Quaternary geology.* New York.

FORBES, E. 1846. On the connection between the distribution of the existing fauna and flora of the British Isles and the geological changes which have affected their area, especially during the epoch of the Northern Drift. *Mem. Geol. Surv. Gt. Br.* Vol. **I.**

GALLOWAY, R. W. 1961a. Periglacial phenomena in Scotland. *Geogr. Annlr.* **43,** 348–53.

1961b. Solifluction in Scotland. *Scott. Geogr. Mag.* **77,** 75–87.

GEIKIE, A. 1863. On the phenomena of the glacial drift in Scotland. *Trans. Geol. Soc. Glasgow* **1,** 1–190.

1865. *The scenery of Scotland.* London.

1887. 2nd ed.

1901. 3rd ed.

GEIKIE, J. 1869. Explanation of sheet 24-Peeblesshire with part of Lanark, Edinburgh and Selkirk. *Mem. Geol. Surv. Scott.,* 1–24.

1873. On the glacial phenomena of the Long Island on Outer Hebrides. *Q. J. Geol. Soc. Lond.* **29,** 532–45.

1874. *The great ice age and its relation to the antiquity of man.* London.

1877. 2nd ed.

1899. 3rd ed.

1878. On the glacial phenomena of the Long Island on Outer Hebrides. *Q. J. Geol. Soc. Lond.* **34,** 819–66.

GEMMELL, A. M. D. 1973. The deglaciation of the island of Arran, Scotland. *Trans. Inst. Br. Geogr.* **59,** 25–39.

GILLBERG, G. 1956. Den glaciala utvecklingen inom sydsvenska hoglandets vastrarandzon issjar och isavsmaltning. *Geol. Foren. Stockh. Forh.* **78,** 357–458.

GJESSING, J. 1960. *The drainage of the deglaciation period, its trends and morphogenetic activity in Northern Atnedalen—with comparitive studies from Northern Gudbrandsdalen and Northern Østerdalen.* Ad Novas 3, Oslo.

GODARD, A. 1965. *Recherches de géomorphologie en Ecosse du Nord-Ouest.* Paris.

Godwin, H. 1940. Pollen analysis and forest history of England and Wales. *New Phytol.* **39,** 370–400.

1975. *The history of the British flora.* Cambridge (2nd ed.).

Godwin, H. & Willis, E. H. 1959. Cambridge University natural radiocarbon measurements. *Radiocarbon* **1,** 63–75.

Goldthwaite, R. P. 1971. *Till—a symposium.* Columbus.

Goodlet, G. A. 1964. The kamiform deposits near Carstairs, Lanarkshire. *Bull. Geol. Surv. Gt. Br.* **21,** 175–96.

Gray, J. M. 1972. The Inter-, Late- and Postglacial shorelines, and ice-limits of Lorn and Eastern Mull. *Univ. of Edinburgh Ph.D. thesis.* Unpubl.

1974a. The main rock platform of the Firth of Lorn, western Scotland. *Trans. Inst. Br. Geogr.* **61,** 81–99.

1974b. Lateglacial and Postglacial shorelines in western Scotland. *Boreas* **3,** 129–38.

1975a. Measurement and analysis of Scottish raised shoreline altitudes. *Dept. of Geogr., Queen Mary College, Occ. Paper* 2.

1975b. Some comments on the Lateglacial history of the Firth of Clyde. *Trans. Inst. Br. Geogr.* **64,** 129–32.

1975c. The Loch Lomond Readvance and contemporaneous sea-levels in Loch Etive and neighbouring areas of western Scotland. *Proc. Geol. Ass.* **86,** 227–38.

1978. Low-level shore platforms in the south-west Scottish Highlands: altitude, age and correlation. *Trans. Inst. Br. Geogr.* New Series **3,** 151–64.

Gray, J. M. & Brooks, C. L. 1972. The Loch Lomond Readvance Moraines of Mull and Menteith. *Scott. J. Geol.* **8,** 95–103.

Gray, J. M. & Lowe, J. J. 1977. The Scottish Lateglacial environment: a synthesis. In Gray, J. M. & Lowe, J. J. (eds.) *Studies in the Scottish Lateglacial environment,* 163–82. Oxford.

Gray, J. M. & Sutherland, D. G. 1977. The "Oban-Ford moraine": a reappraisal. In Gray, J. M. & Lowe, J. J. (eds.) *Studies in the Scottish Lateglacial environment,* 33–44. Oxford.

Gregory, J. C. 1926. The parallel roads of Loch Tulla. *Trans. Geol. Soc. Glasgow* **17,** 91–104.

Gregory, J. W. 1913. The Polmont kame and on the classification of Scottish kames. *Trans. Geol. Soc. Glasgow* **14,** 199–218.

1915a. The kames of Carstairs. *Scott. Geogr. Mag.* **31,** 465–76.

1915b. The Tweed Valley and its relation to the Clyde and Solway. *Scott. Geogr. Mag.* **31,** 478–86.

1926. The Scottish kames and their evidence on the glaciation of Scotland. *Trans. R. Soc. Edinb.* **54,** 392–432.

1927. The moraines, boulder clay and glacial sequence of south-western Scotland. *Trans. Geol. Soc.* Glasgow **17,** 354–70.

Grigson, C. 1966. Animal remains from Fussell's Lodge, Long Barrow. *Archaeologia* **100,** 63–73.

Hall, C. 1812. On revolutions of the earth's surface. *Trans. R. Soc. Edinb.* **7,** 139–212.

Hanson, W. S. & Maciness, L. 1980. Forests, forts and fields: a discussion. *Scottish Arch. Forum* **12,** 98–113.

Harding, R. J. 1978. The variation of the altitudinal gradient of temperature within the British Isles. *Geogr. Annlr.* **60A,** 43–50.

Harkness, D. D. & Wilson, H. W. 1974. Scottish universities research and reactor centre radiocarbon measurements II. *Radiocarbon* **16,** 238–51.

Haynes, U. 1968. The influence of glacial erosion and rock structure on corries in Scotland. *Geogr. Annlr.* **50A,** 221–34.

Hays, J. D. *et al* 1976. Variations in the earth's orbit: pacemaker of the Ice Ages. *Science* **194,** 1121–32.

Hazel, J. E. 1967. Classification and distribution of the recent Hemicytheridae and Trachyceberididae (ostracoda) of north-eastern North America. *U.S. Geol. Surv., Prof. Pap.* 564, 1–49.

1970. Ostracode zoogeography in southern Nova Scotian and northern Virginian faunal provinces. *U.S. Geol. Surv., Prof. Pap.* 29-E, 1–21.

Heim, J. 1962. Recherches sur les relations entre la végétation actuelle et le spectre pollinique récent dans les Ardennes Belges. *Bull. Soc. Roy. Bot. Belgique* **96,** 5–92.

Hibbert, F. A. and Switsur, V. R. 1976. Radiocarbon dating of Flandrian pollen zones in Wales and northern England. *New Phytol.* **77,** 793–807.

Hibbert, J. 1830. Direction of the diluvial

current in Shetland. *Edinb. Jour. Sci.* New series **4,** 85.

HOLDEN, W. G. 1977. The glaciation of central Ayrshire. *Univ. of Glasgow Ph.D. thesis.* Unpubl.

HOLLIN, J. T. 1962. On the glacial history of Antarctica. *J. Glaciol.* **4,** 173–95.

HOLMES, R. 1977. Quaternary deposits of the central North Sea V. The Quaternary geology of the U.K. sector of the North Sea between 56° and 58°N. *Rep. Inst. Geol. Sci. Lond.* 77/14.

HOPPE, G. 1950. Nagra exempel pa glacifluvial dranering fran det inre Norrbotten. *Geogr. Annlr.* **32,** 37–59.

1957. Problems of glacial morphology and the ice age. *Geogr. Annlr.* **39,** 1–18.

1965. Submarine peat in the Shetland Islands. *Geogr. Annlr.* **47A,** 195–203.

1970. The Wurm ice-sheets of northern and arctic Europe. *Arcta. Geogr. Lodziensia* **24,** 205–15.

1974. The glacial history of the Shetland Islands. *Inst. Br. Geogr. Spec. Publ.* **7,** 197–210.

HOPPE, G. *et al* 1965. Fran falt och forskning naturgeografi vid Stockholms Universitet. *Ymer,* 109–28.

HUGHES, M. J. *et al* 1977. Late Quaternary foraminifera and dinoflagellate cysts from boreholes in the U.K. sector of the North Sea between 56° and 58°N. *Rep. Inst. Geol. Sci. Lond.* 77/14, 36–46.

HUGHES, T. *et al* 1977. Was there a late Wurm Arctic ice sheet? *Nature* **266,** 596–602.

IMBRIE, J. I. & IMBRIE, K. P. 1979. *Ice Ages—solving the mystery.* London.

IMBRIE, J. & KIPP, N. G. 1971. A new micropalaeontological method for quantitative palaeontology: application to a Late Pleistocene Caribbean core. In Turekian, K. K. *Late Cenozoic glacial ages.* New Haven.

IMRIE, C. 1812. On the dressed rocks of the Campsie Hills. *Wern. Mem.* **2,** 24.

IVERSEN, J. 1944. Viscum, Hedera and Ilex as climatic indicators. *Geol. Foren. Stockh. Forh.* **66,** 463–83.

1949. The influence of prehistoric man on vegetation. *Danm. Geol. Unders. Ser. 4.3 (6),* 1–25.

1954. The Lateglacial flora of Denmark and its relation to climate and soil. *Danm. Geol. Unders. Ser.* II, 80, 87–119.

1956. Forest clearance in the Stone Age. *Scient. Am.* **194,** 36–41.

1969. Retrogressive development of a forest ecosystem demonstrated by pollen diagrams from a fossil mor. *Oikos. Suppl.* **12,** 35–49.

1973. The development of Denmark's nature since the last glacial. *Danm. Geol. Unders. Ser.* 5, 7-C, 1–126.

JAMIESON, T. F. 1860. On the drift and rolled gravel of Aberdeenshire. *Q. J. Geol. Soc. Lond.* **16,** 347.

JAMIESON, T. F. 1862. On the ice-worm rocks of Scotland. *Q. J. Geol. Soc. Lond.* **18,** 164–84.

1865. On the history of the last geological changes in Scotland. *Q. J. Geol. Soc. Lond.* **21,** 161–203.

JARDINE, W. G. 1962. Post-glacial sediments at Girvan, Ayrshire. *Trans. Geol. Soc. Glasgow* **24,** 262–78.

1963. Pleistocene sediments at Girvan, Ayrshire. *Trans. Geol. Soc. Glasgow* **25,** 4–16.

1964. Post-glacial sea levels in south-west Scotland. *Scott. Geogr. Mag.* **80,** 5–11.

1967a. Sediments of the Flandrian transgression in south-west Scotland: terminology and criteria for facies distinction. *Scott. J. Geol.* **3,** 221–26.

1967b. The Post-glacial marine transgression at Girvan, Ayrshire. *Trans. Inst. Br. Geogr.* **42,** 181–2.

1968. The "Perth" readvance. *Scott. J. Geol.* **4,** 185–7.

1971. Form and age of Late Quaternary shorelines and coastal deposits of south-west Scotland: critical data. *Quaternaria* **14,** 103–14.

1973. The Quaternary geology of the Glasgow District. In Bluck, J. (ed.) *Excursion guide to the geology of the Glasgow District,* 156–69. Geol. Soc. Glasgow.

1975. Chronology of Holocene marine transgression and regression in south-western Scotland. *Boreas* **4,** 173–96.

1976. Some problems in plotting the mean surface level of the North Sea and the Irish Sea during the last 15,000 years. *Geol. Foren. Stockh. Forh.* **98,** 78–82.

1977. The Quaternary marine record in

south-west Scotland and the Scottish Hebrides. *Geol. J. Spec. Issue* **7,** 99–118.

1979. The western (U.K.) shore of the North Sea in Late Pleistocene and Holocene times. In Oele, E. *et al The Quaternary history of the North Sea.* Uppsala.

1980. Holocene raised coastal sediments and former shorelines of Dumfriesshire and eastern Galloway. *Trans. J. Proc. Dumfries. Galloway Nat. Hist. Antiq. Soc.* **55,** 1–59.

1981a. Holocene shorelines in Britain: recent studies. In van Loon, A. J. (ed.) *Quaternary geology: a farewell to A. J. Wigger. Geol. Mijnbouw* **60,** 297–304.

1981b. Status and relationships of the Loch Lomond Readvance and its stratigraphical correlatives. In Neal, J. & Flenley, J. (eds.) *The Quaternary in Britain,* 168–73. Oxford.

1982. Sea level changes in Scotland during the last 18,000 years. *Proc. Geol. Ass.* **93,** 25–41.

JOHANSEN, J. 1975. Pollen diagrams from the Shetland and Faroe Islands. *New Phytol.* **75,** 369–87.

JONASSEN, H. 1950. Recent pollen sedimentation and Jutland heath diagram. *Dansk. Botanisk. Arkiu.* **13**(7), 1–51.

JONES, R. L. 1976. The activities of Mesolithic man: further palaeobotanical evidence from north-east Yorkshire. In Davidson, D. A. & Shackley, M. L. (eds.) *Geo-archaeology: earth science and the past,* 355–67. London.

KEATINGE, T. H. & DICKSON, J. H. 1978. Mid-Flandrian changes in vegetation on Mainland Orkney. *New Phytol.* **82,** 585–612.

KEIHLACK, K. 1926. Das Quartar. In Salmon (ed.) *Grundzuge der geologie,* Vol. **2.** Stuttgart.

KELLOGG, T. B. 1976. Late Quaternary climatic changes: evidence from cores of Norwegian and Greenland Seas. In Cline, R. M. & Hays, J. D. (eds.) *Investigations of Late Quaternary palaeo-oceanography and palaeoclimatology.* Geol. Soc. Am. Mem. **145,** 77–110.

1980. Palaeoclimatology and palaeo-oceanography of the Norwegian and Greenland Seas: glacial-interglacial contrasts. *Boreas* **9,** 115–37.

KELLOGG, T. B. *et al* 1978. Planktonic foraminiferal and oxygen isotopic stratigraphy and palaeoclimatology of Norwegian Sea deep sea cores. *Boreas* **7,** 61–73.

KENDALL, P. F. & BAILEY, E. B. 1908. The glaciation of East Lothian south of the Garleton Hills. *Trans. R. Soc. Edinb.* **46,** 1–31.

KENT, D. *et al* 1971. Climatic change in the North Pacific using ice-rafted detritus as a climatic indicator. *Bull. Geol. Soc. Am.* **82,** 2741–54.

KING, R. B. 1971. Boulder polygons and stripes in the Cairngorm Mountains, Scotland. *J. Glaciol.* **10,** 375–86.

KIRBY, R. P. 1968. The ground moraines of Midlothian and East Lothian. *Scott. J. Geol.* **4,** 209–20.

KIRK, W. & GODWIN, H. 1963. A Late-glacial site at Loch Droma, Ross and Cromarty. *Trans. R. Soc. Edinb.* **65,** 225–49.

KOPPEN, W. & WAGNER, A. 1924. *Die klimate der geologischen vorzeit.* Berlin.

KURTEN, B. 1968. *Pleistocene mammals of Europe.* London.

LACAILLE, A. D. 1954. *The Stone Age in Scotland.* Oxford.

LAMB, H. H. 1972. *Climate: past, present and future*—vol. **1.** London.

1977. *Climate: past, present and future*—vol. **2.** *Climatic history and the future.* London.

LAWSON, T. J. 1981. The first Scottish date from the last Interglacial. *Scott. J. Geol.* **17,** 301–03.

LEDGER, D. C. *et al* 1980. Rate of sedimentation in Kelly Reservoir, Strathclyde. *Scott. J. Geol.* **16,** 281–85.

LEVERETT, F. 1898. The weathered zone (Sangamon) between the Iowan loess and Illinoian till. *J. Geol.* **6,** 171–81.

1899. The Illinois glacial lobe. *U.S. Geol. Surv. Mon.* **38,** 817 pp.

LEWIS, F. J. 1905. The plant remains in the Scottish peat mosses, Part I. The Scottish Southern Uplands. *Trans. R. Soc. Edinb.* **41,** 699–723.

1906. The plant remains in the Scottish peat mosses, Part II. The Scottish Higlands. *Trans. R. Soc. Edinb.* **45,** 335–60.

1907. The plant remains in the Scottish peat mosses, Part III. The Scottish Highlands and the Shetland Islands. *Trans. R. Soc. Edinb.* **46,** 33–70.

1911. The plant remains in the Scottish peat mosses, Part IV. The Scottish Highlands and Shetland with an appendix on the Icelandic peat deposits. *Trans. R. Soc. Edinb.* **47,** 793–833.

LIDZ, L. 1966. Deep sea biostratigraphy. *Science* **154,** 1448.

LINTON, D. L. 1949. Some Scottish river captures re-examined. *Scott. Geogr. Mag.* **65,** 123–32.

1951a. Problems of Scottish scenery. *Scott. Geogr. Mag.* **67,** 65–85.

1951b. Watershed breaching by ice in Scotland. *Trans. Inst. Br. Geogr.* **15,** 1–15.

1959. Morphological contrasts between eastern and western Scotland. In Miller, R. & Watson, J. W. (eds.) *Geographical essays in memory of Alan G. Ogilvie,* 16–45. Edinburgh.

1963. The forms of glacial erosion. *Trans. Inst. Br. Geogr.* **33,** 1–28.

LORD, A. R. 1980. Interpretation of the Lateglacial marine environment of north-west Europe by means of foraminifera. In Lowe, J. J., Gray, J. M. & Robinson, J. E. (eds.) *Studies in the Lateglacial of north-west Europe,* 103–114. Oxford.

LOVELL, J. P. B. *et al* 1973. Rate of sedimentation in the North Esk Reservoir, Midlothian. *Scott. J. Geol.* **9**(1)**,** 57–62.

LOWE, J. J. & GRAY, J. M. 1980. The stratigraphic subdivision of the Lateglacial of N.W. Europe. In Lowe, J. J., Gray, J. M. & Robinson, J. E. (eds.) *Studies in the Lateglacial of north-west Europe,* 157–76. Oxford.

LOWE, J. J. & WALKER, M. J. C. 1977. The reconstruction of the Lateglacial environment in the southern and eastern Grampian Highlands. In Gray, J. M. & Lowe, J. J. (eds.) *Studies in the Scottish Lateglacial environment,* 101–18. Oxford.

1980a. Problems associated with radiocarbon dating the close of the Lateglacial period in the Rannoch Moor area, Scotland. In Lowe, J. J., Gray, J. M. & Robinson, J. E. (eds.) *Studies in the Lateglacial of north-west Europe,* 123–38. Oxford.

1980b. The early Postglacial environment of Scotland: evidence from a site near Tyndrum, Perthsire. *Boreas* **10,** 281–94.

LOWE, J. J. *et al* (eds.) 1980. *Studies in the Lateglacial of North-west Europe.* Oxford.

LYTHE, S. G. & BUTT, J. 1975. *An Economic History of Scotland 1100–1939.* Glasgow.

MCCALLIEN, W. J. 1937. Lateglacial and early Postglacial Scotland. *Proc. Soc. Antiq. Scot.* **71,** 174–206.

MCCANN, S. B. 1961a. Some supposed "raised beach" deposits at Corran, Loch Linnhe and Loch Etive. *Geol. Mag.* **98,** 131–42.

1961b. The raised beaches of western Scotland. *Univ. of Cambridge Ph.D. thesis.* Unpubl.

1964. The raised beaches of north-east Islay and western Jura, Argyll. *Trans. Inst. Br. Geogr.* **35,** 1–16.

1966a. The limits of the Lateglacial Highland, or Loch Lomond, Readvance along the west Highland seaboard from Oban to Mallaig. *Scott. J. Geol.* **2,** 84–95.

1966b. The main Postglacial raised shoreline of western Scotland, from the Firth of Lorne to Loch Broom. *Trans. Inst. Br. Geogr.* **39,** 87–99.

1968. Raised rock platforms in the Western Isles of Scotland. In Bowen, E. G. *et al* (eds.) *Geography at Aberystwyth,* 22–34. London.

MCCANN, S. B. & RICHARDS, A. 1969. The coastal features of the Island of Rhum in the Inner Hebrides. *Scott. J. Geol.* **5,** 15–25.

MACCULLOCH, A. 1812. On a perched block near Dunkeld. *Edinb. J. Sci* **3,** 46.

MCGREGOR, M. 1927. The Carstairs district. *Proc. Geol. Ass.* **36,** 495–99.

MCGREGOR, M. & MCGREGOR, A. G. 1936. The Midland Valley of Scotland. *Br. Reg. Geol.* London.

MCINTYRE, A. & RUDDIMAN, W. F. 1972. North-east Atlantic Post-eemian palaeo-oceanography; a predictive analogue of the future. *Quat. Res.* **2,** 350–54.

MCINTYRE, A. *et al* 1972. Southward penetrations of the North Atlantic polar front. *Deep-Sea Research* **19,** 61–77.

1976. Glacial North Atlantic 18,000 years ago. A Climap reconstruction. In Cline, R. M. & Hays, J. D. (eds.) Investigations of Late Quaternary palaeo-oceanography and palaeoclimatology. *Geol. Soc. Am. Mem.* **145,** 43–76.

MACKIE, E. W. 1972. Radiocarbon dates for two Mesolithic shell heaps and a Neolithic axe factory in Scotland. *Proc. Prehist. Soc.* **38,** 412–16.

MACKINDER, H. 1915. *Britain and the British seas.* Oxford.

McLellan, A. G. 1967a. The distribution, origin and use of sand and gravel deposits in central Lanarkshire. *Univ. of Glasgow Ph.D. thesis*. Unpubl.

1967b. *The distribution of sand and gravel deposits in west central Scotland and some problems concerning their utilisation*. Glasgow University.

1969. The last glaciation and deglaciation of central Lanarkshire. *Scott. J. Geol.* **5,** 248–68.

1970. Sand and gravel in Scotland. *Cement, Lime and Gravel,* 154–56.

MacPherson, J. B. 1980. Environmental change during the Loch Lomond Stadial: evidence from a site in the Upper Spey Valley, Scotland. In Lowe, J. J. *et al* (eds.) *Studies in the Lateglacial of north-west Europe*, 89–102. Oxford.

McVean, D. N. & Ratcliffe, D. A. 1962. *Plant communities of the Scottish Highlands*. Nature Con. Mon. No. 1. H.M.S.O. London.

Mangerud, J. *et al* 1974. Quaternary stratigraphy of Norden, a proposal for terminology and classification. *Boreas* **3,** 109–27.

Mannion, A. M. 1978. A palaeogeographical study from south-east Scotland. *Univ. of Reading Geogr. Papers* 67, 42 pp.

Manley, G. 1951. The range of variation of the British climate. *Geogr. J.* **117,** 43–68.

Mannerfelt, C. M. 1945. Nagra glacialmorfologiska formelement. *Geogr. Annlr.* **27,** 1–239.

1949. Marginal drainage channels as indicators of the gradients of Quaternary ice caps. *Geogr. Annlr.* **31,** 194–9.

Mathieson, J. & Bailey, E. B. 1925. The glacial strandlines at Loch Tulla. *Trans. Edinb. Geol. Soc.* **11,** 193–9.

May, J. 1981. The glaciation and deglaciation of upper Nithsdale and Annandale. *Univ. of Glasgow Ph.D. thesis*. Unpubl.

Mellars, P. A. 1976. Fire ecology, animal populations and man: a study of some ecological relationships in prehistory. *Proc. Prehist. Soc.* **42,** 15–45.

1979. Excavation and economic analysis of Mesolithic shell midens on the island of Oronsay. In Thoms, L. M. (ed.) Early man in the Scottish landscape: *Scottish Archaeological Forum* **9,** 43–61.

Mellars, P. A. & Payne, S. 1971. Excavation of two Mesolithic shell midens on the island of Oronsay. *Nature* **231,** 397–8.

Menzies, J. 1976. The glacial geomorphology of Glasgow with particular reference to the drumlins. *Univ. of Edinb. Ph.D. thesis*. Unpubl.

1981. Investigations into the Quaternary deposits and bedrock topography of central Glasgow. *Scott. J. Geol.* **17,** 155–68.

Mercer, J. 1968. Sone tools from a washing-limit deposit of the highest Post-glacial transgression, Lealt Bay, Isle of Jura. *Proc. Soc. Antiq. Scot.* **100,** 1–46.

1970a. Flint tools from the present tidal zone, Lussa Bay, Isle of Jura, Argyll. *Proc. Soc. Antiq. Scot.* **102,** 1–30.

1970b. The microlithic succession in North Jura, Argyll, west Scotland. *Quaternaria* **13,** 177–85.

1971. A regression-time stone-workers' camp 33 feet O.D., Lussa River, Isle of Jura. *Proc. Soc. Antiq. Scot.* **103,** 1–32.

1972. Microlithic and Bronze Age camps, 75–62 feet O.D., N. Carn, Isle of Jura. *Proc. Soc. Antiq. Scot.* **104,** 1–22.

1974a. New Cl4 dates from the Isle of Jura. *Antiquity* **48,** 65–6.

1974b. *Hebridean Islands: Colonsay, Gigha, Jura*. Glasgow.

1974c. Glenbatrick waterhole, a microlithic site on the Isle of Jura. *Proc. Soc. Antiq. Scot.* **105,** 9–32.

Mercer, J. H. 1969. The Alleröd oscillation: a European climatic anomaly? *Arct. Alp. Res.* **1,** 227–34.

Mesolella, K. J. *et al* 1969. The astronomical theory of climatic change: Barbados data. *J. Geol.* **77,** 250–74.

Milankovitch, M. M. 1941. *Canon of insolation and the ice age problem*. Koniglick Serbische Akademie, Beograd. (English translation by the Israel programme for scientific translations. Published for the U.S. Dept. of Commerce and the N.S.F., Washington D.C.).

Miller, R. *et al* 1954. Stone stripes and other surface features of Tinto Hill. *Geogr. J.* **120,** 216–19.

Milne, D. 1847. On the parallel roads of Loch Tulla. *Trans. Geol. Soc. Glasgow* **17,** 91–104.

Moar, N. T. 1969a. Late Weichselian and Flandrian pollen diagrams from south-west Scotland. *New Phytol.* **68,** 433–67.

1969b. A radiocarbon-dated pollen diagram from north-west Scotland. *New Phytol.* **68,** 209–14.

1969c. Two pollen diagrams from the Mainland, Orkney Islands. *New Phytol.* **68,** 201–8.

MOORE, P. D. 1980. The reconstruction of the Lateglacial environment: some problems associated with the interpretation of pollen data. In Lowe, J. J., Gray, J. M. & Robinson, J. E. (eds.) *Studies in the Lateglacial of north-west Europe,* 151–56. Oxford.

MORRISON, A. 1980. *Early man in Britain and Ireland.* London.

MORRISON, J. *et al* 1981. The culmination of the main Postglacial transgression in the Firth of Tay area, Scotland. *Proc. Geol. Ass.* **92,** 197–209.

NEWEY, W. W. 1965. Post-glacial vegetational and climatic changes in part of south-east Scotland. *Univ. of Edinb. Ph.D. thesis.* Unpubl.

1966. Pollen analysis of sub-carse peats of the Forth Valley. *Trans. Inst. Br. Geogr.* **39,** 53–9.

1967. Pollen analyses from south-east Scotland. *Trans. Bot. Soc. Edinb.* **40,** 424–34.

1970. Pollen analysis of Late-Weichselian deposits at Corstorphine, Edinburgh. *New Phytol.* **69,** 1167–77.

NICHOLS, H. 1967. Vegetational change, shoreline displacement and the human factor in the Late Quaternary history of south-west Scotland. *Trans. R. Soc. Edinb.* **67,** 145–87.

NILSSON, T. 1948. On the application of the Scanian Post-glacial zone system to the Danish pollen diagrams. *K. Danske Vidensk. Selsk. Biol. Skr.* **5,** 1–53.

O'DELL, A. C. & WALTON, K. 1962. *The Highlands and Islands of Scotland.* London.

OLDFIELD, F. 1960. Studies in the Post-glacial history of British vegetation: Lowland Lonsdale. *New Phytol* **59,** 192–217.

O'SULLIVAN, P. E. 1974a. Two Flandrian pollen diagrams from the east central Highlands of Scotland. *Pollen Spores* **16,** 33–57.

1974b. Radiocarbon dating and prehistoric forest clearance on Speyside. *Proc. Prehist. Soc.* **40,** 206–8.

1976. Pollen analysis and radiocarbon dating of a core from L. Pityoulish, East Highlands, Scotland. *J. Biogeog.* **3,** 293–302.

OWENS, R. 1977. Quaternary deposits of the central North Sea. Preliminary report on the superficial sediments of the central North Sea between 56° and 59°N. and 2°W. and 2°E. *Rep. Inst. Geol. Sci. Lond.* 73/13.

PAGE, N. R. 1972. On the age of the Hoxnian Interglacial. *Geol. J.* **8,** 129–42.

PANTIN, H. M. 1975. Quaternary sediments of the north-eastern Irish Sea. *Quaternary Newsletter* **17,** 7–9.

PARRY, M. L. 1975. Secular climatic change and marginal land. *Trans. Inst. Br. Geogr.* **64,** 1–13.

1976. Abandoned farmland in upland Britain. *Geogr. J.* **142,** 101–10.

1978. *Climatic change agriculture and settlement.* Folkestone.

PARRY, M. L. & SLATER, T. R. (eds.) 1980. *The making of the Scottish countryside.* London.

PATERSON, I. B. 1974. The supposed Perth Readvance in the Perth district. *Scott. J. Geol.* **10,** 53–66.

1981. The Quaternary geology of the Buddon Ness area of Tayside, Scotland. *Inst. Geol. Sci. Rep.* 81/1.

PATERSON, I. B. *et al* 1981. Quaternary estuarine deposits in the Tay-Earn area, Scotland. *Inst. Geol. Sci. Rep.* 81/7.

PEACH, A. M. 1909. Boulder distribution from Lennoxtown, Scotland. *Geol. Mag.* **46,** 26–31.

PEACH, B. N. & HORNE, J. 1879. The glaciation of the Shetland Isles. *Q. J. Geol. Soc. Lond.* **35,** 778–811.

PEACOCK, J. D. 1970. Some aspects of the glacial geology of west Inverness-shire. *Bull. Geol. Surv. Gt. Br.* **33,** 125–45.

1971. Marine shell radiocarbon dates and the chronology of deglaciation in western Scotland. *Nature Phys. Sci.* **230,** 43–5.

1974. Borehole evidence for Late- and Post-glacial events in the Cromarty Firth, Scotland. *Bull. Geol. Surv. Gt. Br.* **48,** 55–67.

1975. Scottish Late- and Post-glacial marine deposits. In Gemmell, A. M. D. (ed.) *Quaternary Studies in north east Scotland,* 45–8. Aberdeen.

1981. Scottish Lateglacial marine deposits and their environmental significance. In Neal, J. & Flenley, J. (eds.) *The Quaternary in Britain.* Oxford.

PEACOCK, J. D. *et al* 1977. Evolution and chronology of Lateglacial marine environments at Lochgilphead, Scotland. In Gray, J. M. &

Lowe, J. J. (eds.) *Studies in the Scottish Lateglacial environment,* 89–100. Oxford.

1978. Lateglacial and Postglacial marine environments at Ardyne, Scotland and their significance in the interpretation of the history of the Clyde Sea area. *Inst. Geol. Sci. Rep.* 78/17.

1980. Late- and Post-glacial marine environments in part of the Inner Cromarty Firth, Scotland. *Inst. Geol. Sci. Rep.* 80/7.

Peacock, J. D. & Ross, D. L. 1978. Anomalous glacial erratics in the southern part of the Outer Hebrides. *Scott. J. Geol.* **14,** 263.

Pears, N. U. 1968. Postglacial tree-lines of the Cairngorm Mountains, Scotland. *Trans. Bot. Soc. Edinb.* **40,** 361–94.

1972. Interpretation problems in the study of treeline fluctuations. In Taylor, J. A. (ed.) *Research papers in forest meteorology, an Aberystwyth symposium,* 31–45. Aberystwyth.

Peglar, S. 1979. A radiocarbon dated pollen diagram from Loch of Winless, Caithness. *New Phytol.* **82,** 245–63.

Penck, A. & Bruckner, E. 1909. *Die Alpen in Eiszeitalter.* Leipzig.

Pennington, W. 1975. A chronostratigraphic comparison of Late Weichselian and Late Devensian subdivisions, illustrated by two radiocarbon-dated profiles from western Britain. *Boreas* **4,** 157–71.

1977a. The Late Devensian flora and vegetation of Britain. *Phil. Trans. R. Soc. Lond.* B. **280,** 247–71.

1977b. Lake sediments and the Lateglacial environment in northern Scotland. In Gray, J. M. & Lowe, J. J. (eds.) *Studies in the Scottish Lateglacial Environment,* 119–42. Oxford.

Pennington, W. & Lishman, J. P. 1971. Iodine in lake sediments in northern England and Scotland. *Biol. Rev.* **46,** 279–313.

Pennington, W. *et al* 1972. Lake sediments in northern Scotland. *Phil. Trans. R. Soc. Lond.* B. **264,** 191–294.

Péwé, T. L. 1966. Palaeoclimatic significance of fossil ice wedges. *Biul. Peryglac.* **15,** 65–73.

Piggot, S. 1965. *Ancient Europe.* Edinburgh.

Playfair, J. 1802. *Illustrations of the Huttonian theory of the earth.* Edinburgh.

Post, A. 1976. The tilted forest: glaciological-geological implications of vegetated neoglacial ice at Lituya Bay, Alaska. *Quat. Res.* **6,** 111–18.

Price, R. J. 1960. Glacial meltwater channels in the upper Tweed drainage basin. *Geogr. J.* **126,** 485–89.

1961. The deglaciation of the Tweed drainage area west of Innerleithen. *Univ. of Edinburgh Ph.D. thesis.* Unpubl.

1963a. A glacial meltwater drainge system in Peeblesshire, Scotland. *Scott. Geogr. Mag.* **79,** 133–41.

1963b. The glaciation of a part of Peeblesshire, Scotland. *Trans. Edinb. Geol. Soc.* **19,** 326–48.

1965. The changing proglacial environment of the Casement Glacier, Glacier Bay, Alaska. *Trans. Inst. Br. Geogr.* **36,** 107–16.

1966. Eskers near the Casement Glacier, Glacier Bay, Alaska. *Geogr. Annlr.* **48,** 111–25.

1969. Moraines, sandar, kames and eskers near Breidamerkurjokull, Iceland. *Trans. Inst. Br. Geogr.* **46,** 17–43.

1970. Moraines at Fjallsjokull, Iceland. *J. Arct. Alp. Res.* **2,** 27–42.

1971. The development and destruction of a sandur, Breidamerkurjokull, Iceland. *J. Arct. Alp. Res.* **3,** 225–37.

1973. *Glacial and fluvioglacial landforms.* Edinburgh.

1974. The glaciation of west central Scotland. A review. *Scott. Geogr. Mag.* **91,** 131–45.

1976. *Highland landforms.* Glasgow.

1980a. Rates of geomorphological changes in proglacial areas. In Cullingford, R. A. *et al* (eds.) *Timescales in geomorphology,* 79–93. Chichester.

1980b. Geomorphological implications of environmental changes during the last 30,000 years in central Scotland. *Z. Geomorph.*, Suppl. Bd. **36,** 74–83.

Richards, A. 1969. Some aspects of the evolution of the coastline of Northeast Skye. *Scott. Geogr. Mag.* **85,** 122–31.

Ritchie, J. 1920. *The influence of man on animal life in Scotland.* Cambridge.

Ritchie, W. 1966. The Postglacial rise in sea-level and coastal changes in the Uists. *Trans. Inst. Br. Geogr.* **39,** 79–86.

1972. The evolution of coastal sand dunes. *Scott. Geogr. Mag.* **88,** 19–35.

Robinson, G. *et al* 1971. Trend surface analysis

of corrie altitudes in Scotland. *Scott. Geogr. Mag.* **87,** 142–6.

ROBINSON, J. E. 1980. The marine ostracod record from the Lateglacial period in Britain and Northwest Europe: a review. In Lowe, J. J., Gray, J. M. & Robinson, J. E. (eds.) *Studies in the Lateglacial of Northwest Europe,* 115–22. Oxford.

ROBINSON, M. & BALLANTYNE, C. K. 1979. Evidence for a glacial readvance pre-dating the Loch Lomond Advance in Wester Ross. *Scott. J. Geol.* **15,** 271–77.

ROLFE, W. D. I. 1966. Woolly Rhinoceros from the Scottish Pleistocene. *Scott. J. Geol.* **2,** 253–8.

ROMANS, J. C. C. *et al* 1966. Alpine Soils of north-east Scotland. *J. Soil. Sci.* **17,** 184–99.

ROSE, J. 1975. Raised beach gravels and ice wedge casts at Old Kilpatrick, near Glasgow. *Scott. J. Geol.* **11,** 15–21.

1980. In Jardine, W. G. (ed.) *Quat. Res. Ass. Field Guide—Glasgow Region,* 22–39. Glasgow.

ROSE, J. & LETZER, J. M. 1977. Superimposed drumlins. *J. Glaciol.* **18,** 471–80.

ROSS, G. 1920. *Summ. Prog. Geol. Surv. Gt. Br.,* 158–60.

RUDDIMAN, W. F. & MCINTYRE, A. 1973. Time-transgressive deglacial retreat of polar waters from the North Atlantic. *Quat. Res.* **3,** 117–30.

1976. Northeast Atlantic palaeoclimatic changes over the last 600,000 years. In Cline, R. M. & Hays, J. D. (eds.) *Investigations of Late Quaternary palaeo-oceanography and palaeoclimatology.* Geol. Soc. Am. Mem. **145,** 199–214.

1981a. The mode and mechanism of the last deglaciation: oceanic evidence. *Quat. Res.* **16,** 125–34.

1981b. The North Atlantic Ocean during the last deglaciation. *Palaeogeog. Palaeoclim. Palaeoecol.* **35,** 145–214.

RUDDIMAN, W. F. *et al* 1977. Glacial/interglacial response rate of subpolar North Atlantic waters to climatic change: the record left in deep-sea sediments. *Phil. Trans. R. Soc. Lond.* B. **280,** 119–42.

1980. Oceanic evidence for the mechanism of rapid northern hemisphere glaciation. *Quat. Res.* **13,** 33–64.

RYDER, R. & MCCANN, S. B. 1971. Periglacial phenomena on the island of Rhum in the Inner Hebrides. *Scott. J. Geol.* **7,** 293–303.

RYMER, L. 1974. The palaeoecology and historical ecology of the parish of north Knapdale. *Univ. of Cambridge Ph.D. thesis.* Unpubl.

SAMUELSSON, G. 1910. Scottish peat mosses. A contribution to the knowledge of the Late Quaternary vegetation and climate of northwestern Europe. *Bull. Geol. Inst. Uppsala* **10,** 197–260.

SANCETTA, C., IMBRIE, J. & KIPP, N. G. 1973. Climatic record of the past 130,000 years in North Atlantic deep-sea core V23-82: correlation with the terrestrial record. *Quat. Res.* **3,** 110–16.

SCHOTT, W. 1935. Die formaniferen in dem Aquatorialel teil des Atlantischen Ozeans. *Deutsch. Atlant. Exped. Meteor. 1925–1927.* Wiss., Ergebn **3,** 43–134.

SERNANDER, R. 1908. On the evidence of Postglacial changes of climate furnished by the peat mosses of northern Europe. *Geol. Foren Stockh. Forh.* **30,** 465–78.

SHACKLETON, N. J. & OPDYKE, N. D. 1973. Oxygen isotope and palaeomagnetic stratigraphy of Equatorial Pacific core V28–238: oxygen isotope temperature and ice volumes on a 10^5 and 10^6 year scale. *Quat. Res.* **3,** 39–55.

SHAKESBY, R. A. 1978. Dispersal of glacial erratics from Lennoxtown, Stirlingshire. *Scott. J. Geol.* **14,** 81–86.

SHOTTON, F. W. *et al* 1970. Birmingham University radiocarbon dates IV. *Radiocarbon* **12,** 385–9.

SHOTTON, F. W. 1972. An example of hard water error in radiocarbon dating of vegetable matter. *Nature,* London, **240,** 460–1.

1977. British dating work with radioactive isotopes. In Shotton, F. W. (ed.) *British Quaternary Studies,* 17–29. Oxford.

SIMMONS, I. G. & CUNDILL, P. R. 1974. Late Quaternary vegetational history of the North York Moors. *J. Biogr.* **1,** 253–61.

SIMPSON, I. M. & WEST, R. G. 1958. On the stratigraphy and palaeobotany of a Late Pleistocene organic deposit at Chelford, Cheshire. *New Phytol.* **57,** 239–50.

SIMPSON, J. B. 1933. The Lateglacial readvance moraines of the Highland border west of the River Tay. *Trans. R. Soc. Edinb.* **57,** 633–45.

SISSONS, J. B. 1958a. The deglaciation of part of East Lothian. *Trans. Inst. Br. Geogr.* **25,** 59–77.

1958b. Supposed ice-dammed lakes in Britain with particular reference to the Eddleston Valley, southern Scotland. *Geogr. Annlr.* **40,** 159–87.

1958c. Subglacial stream erosion in southern Northumberland. *Scott. Geogr. Mag.* **74,** 163–74.

1960a. Some aspects of glacial drainage channels in Britain, part I. *Scott. Geogr. Mag.* **76,** 131–46.

1960b. Subglacial, marginal and other glacial drainage in the Syracuse-Oneida areas, New York. *Bull. Geol. Soc. Am.* **71,** 1575–88.

1960c. Erosion surfaces, cyclic slopes and drainage systems in southern Scotland and northern England. *Trans. Inst. Br. Geogr.* **28,** 23–38.

1961a. Some aspects of glacial drainage channels in Britain, part II. *Scott. Geogr. Mag.* **77,** 15–36.

1961b. A sublacial drainage system by the Tinto Hills, Lanarkshire. *Trans. Edinb. Geol. Soc.* **18,** 175–92.

1961c. The central and eastern parts of the Lammermuir-Stranraer moraine. *Geol. Mag.* **98,** 380–92.

1962. A reinterpretation of the literature on Lateglacial shorelines in Scotland with particular reference to the Forth area. *Trans. Edinb. Geol. Soc.* **19,** 83–99.

1963a. The glacial drainage system around Carlops, Peeblesshire. *Trans. Inst. Br. Geogr.* **32,** 95–111.

1963b. Scottish raised shoreline heights with particular reference to the Forth Valley. *Geogr. Annlr.* **45,** 180–5.

1964. The Perth readvance in central Scotland. *Scott. Geogr. Mag.* **79,** 151–63 (part I); **80,** 28–36 (part II).

1965. Quaternary. In Craig, G. Y. (ed.) *The geology of Scotland.* Edinburgh.

1966. Relative sea-level changes between 10,300 and 8,300 bp in part of the carse of Stirling. *Trans. Inst. Br. Geogr.* **39,** 19–29.

1967a. *The evolution of Scotland' scenery.* Edinburgh.

1967b. Comments on the paper by F. M. Synge & N. Stephens in Transactions no. 39. *Trans. Inst. Br. Geogr.* **42,** 163–8.

1967c. Glacial stages and radiocarbon dates in Scotland. *Scott. J. Geol.* **3,** 375–81.

1968. The 'Perth' readvance. *Scott. J. Geol.* **4,** 186–7.

1969. Drift stratigraphy and buried morphological features in the Grangemouth-Falkirk-Airth area, central Scotland. *Trans. Inst. Br. Geogr.* **48,** 19–50.

1971. The geomorphology of central Edinburgh. *Scott. Geogr. Mag.* **87,** 185–96.

1972a. The last glaciers in part of the south-east Grampians. *Scott. Geogr. Mag.* **88,** 168–81.

1972b. Dislocation and non-uniform uplift of raised shorelines in the western part of the Forth Valley. *Trans. Inst. Br. Geogr.* **55,** 145–59.

1973a. Hypothesis of deglaciation in the eastern Grampians, Scotland. *Scott. J. Geol.* **9,** 96.

1973b. Delimiting the Loch Lomond Readvance in the eastern Grampians. *Scott. Geogr. Mag.* **89,** 138–9.

1974a. The Quaternary in Scotland: a review. *Scott. J. Geol.* **10,** 311–38.

1974b. Glacial readvances in Scotland. In Caseldine, C. & Mitchell, W. A. (eds.) *Problems of the deglaciation of Scotland,* 5–15. St. Andrews.

1974c. A Lateglacial ice cap in the central Grampians, Scotland. *Trans. Inst. Br. Geogr.* **62,** 95–114.

1974d. Lateglacial marine erosion in Scotland. *Boreas* **3,** 41–48.

1975a. The Loch Lomond Readvance in the south-east Grampians. In Gemmell, A. M. D. (ed.) *Quaternary studies in north-east Scotland,* 23–9. Aberdeen.

1975b. A fossil rock glacier in Wester Ross. *Scott. J. Geol.* **11,** 83–6.

1976a. *The geomorphology of the British Isles: Scotland.* London.

1976b. A remarkable protalus rampart in Wester Ross. *Scott. Geogr. Mag.* **92,** 182–90.

1976c. Lateglacial marine erosion in south-east Scotland. *Scott. Geogr. Mag.* **92,** 17–29.

1977a. The Loch Lomond Readvance in southern Skye and some palaeoclimatic implications. *Scott. J. Geol.* **13,** 23–36.

1977b. The Loch Lomond Readvance in the northern mainland of Scotland. In Gray,

J. M. & Lowe, J. J. (eds.) *Studies in the Lateglacial environment of Scotland,* 45–60. Oxford.

1977c. Former ice-dammed lakes in Glen Moriston Inverness-shire, and their significance in upland Britain. *Trans. Inst. Br. Geogr.* **2,** 224–42.

1977d. *The Scottish Highlands.* Guidebook for excursions All and Cll, *X* I.N.Q.U.A. Congress.

1978. The parallel roads of Glen Roy and adjacent glens, Scotland. *Boreas* **7,** 229–44.

1979a. The Loch Lomond Stadial in the British Isles. *Nature* **280,** 199–203.

1979b. The Loch Lomond Advance in the Cairngorm Mountains. *Scott. Geogr. Mag.* **95,** 66–82.

1979c. The limit of the Loch Lomond Advance in Glen Roy and vicinity. *Scott. J. Geol.* **15,** 31–42.

1979d. Catastrophic lake drainage in Glen Spean and the Great Glen, Scotland. *J. Geol. Soc.* **136,** 215–24.

1979e. Palaeoclimatic inferences from former glaciers in Scotland and the Lake District. *Nature* **278,** 518–21.

1979f. The parallel roads of Glen Roy. *Nature Conservancy Council,* 8 pp.

1980a. The Loch Lomond Advance in the Lake District, northern England. *Trans. R. Soc. Edinb.* **71,** 13–28.

1980b. Palaeoclimatic inferences from Loch Lomond Advance glaciers. In Lowe, J. J., Gray, J. M. & Robinson, J. E. (eds.) *Studies in the Lateglacial of north-west Europe,* 31–44. Oxford.

1980c. The glaciation of the Outer Hebrides. *Scott. J. Geol.* **16.**

1981a. The last Scottish ice-sheet: facts and speculative discussion. *Boreas* **10,** 1–17.

1981b. British shore platforms and ice sheets. *Nature* **291,** 473–75.

1981c. Lateglacial marine erosion and a Jokulhlaup deposit in the Beauly Firth. *Scott. J. Geol.* **17,** 7–20.

Sissons, J. B. & Brooks, C. L. 1971. Dating of the early Postglacial land and sea-level changes in the western Forth Valley. *Nature Phys. Sci.* **234,** 124–7.

Sissons, J. B. & Dawson, A. G. 1981. Former sea-levels and ice limits in part of Wester Ross, north-west Scotland. *Proc. Geol. Ass.* **92,** 115–24.

Sissons, J. B. & Grant, A. J. H. 1972. The last glaciers in the Lochnagar area, Aberdeenshire. *Scott. J. Geol.* **8,** 85–94.

Sissons, J. B. & Rhind, D. W. 1970. Drift stratigraphy and buried morphology beneath the Forth at Rosyth. *Scott. J. Geol.* **6,** 272–84.

Sissons, J. B. & Smith, D. E. 1965a. Raised shorelines associated with the Perth readvance in the Forth Valley and their relation to glacial isostasy. *Trans. R. Soc. Edinb.* **66,** 23–27.

1965b. Peat bogs in a Postglacial sea and a buried raised beach in the western part of the Carse of Stirling. *Scott. J. Geol.* **1,** 247–55.

Sissons, J. B. & Sutherland, D. G. 1976. Climatic inferences from former glaciers in the south-east Grampian Highlands, Scotland. *J. Glaciol.* **17,** 325–46.

Sissons, J. B. & Walker, M. J. C. 1974. Lateglacial site in the central Grampian Highlands. *Nature* **249,** 822–4.

Sissons, J. B. *et al* 1965. Some pre-carse valleys in the Forth and Tay Basins. *Scott. Geogr. Mag.* **81,** 117–24.

1966. Lateglacial and Postglacial shorelines in south-east Scotland. *Trans. Inst. Br. Geogr.* **39,** 9–18.

1973. Loch Lomond Readvance in the Grampian Highlands of Scotland. *Nature Phys. Sci.* **244,** 75–7.

Smith, A. G. 1970. The influence of Mesolithic and Neolithic man on British vegetation: a discussion. In Walker. D. & West, R. G. (eds.) *Studies in the vegetational history of the British Isles: essays in honour of Harry Godwin,* 81–96.

1981. The Neolithic. In Simmons, I. & Tooley, M. (eds.) *The environment in British prehistory,* 129–209. London.

Smith, D. E. 1968. Postglacial displaced shorelines in the surface of the carse clay on the north bank of the River Forth, in Scotland. *Zeits. Fur. Geomorph.* **12,** 388–408.

Smith, D. E. & Cullingford, R. A. 1981. New evidence for Late Devensian marine limits in east-central Scotland—some comments. *Quat. Newsletter* **35,** 12–24.

Smith, D. E. *et al* 1969. Isobases for the Main Perth raised shoreline in south-east Scotland as determined by trend-surface analysis. *Trans. Inst. Br. Geogr.* **46,** 45–52.

1978. The Late Devensian and Flandrian history of the Teith Valley, Scotland. *Boreas* **7,** 97–107.

1980. Dating the Main Postglacial shoreline in the Montrose area, Scotland. In Cullingford, R. A. *et al* (eds.) *Timescales in geomorphology,* 225–45. Chichester.

1982. Flandrian relative sea-level changes in the Philorth Valley, north-east Scotland. *Trans. Inst. Br. Geogr.* **7,** 321–336.

SMITH, J. 1839. On the Newer Tertiary or Pliocene deposits of Scotland. *Proc. R. Soc. Edinb.* **1,** 263.

1898. The drift or glacial deposits of Ayrshire. *Trans. Geol. Soc. Glasgow,* Suppl. 11, 134 pp.

1902. Marine shells. *Geol. Mag.* **39,** 479.

SMITH, J. S. 1977. The last glacial epoch around the Moray Firth. In *The Moray Firth area—geological studies.* Inverness Field Club, 72–82. Inverness.

STEPHEN, D. 1974. *Highland Animals.* Glasgow.

STEWART, M. 1933. Notes on the geology of Sula Sgeir and the Flannan Islands. *Geol. Mag.* **70,** 110–16.

STUART, A. J. 1974. Pleistocene history of British vertebrate fauna. *Biol. Rev.* **49,** 225–66.

1977. British Quaternary vertebrates. In Shotton, F. W. (ed.) *British Quaternary studies—recent advances,* 69–82. Oxford.

SUGDEN, D. E. 1968. The selectivity of glacial erosion in the Cairngorm Mountains. *Trans. Inst. Br. Geogr.* **45,** 79–92.

1969. The age and form of corries in the Cairngorms. *Scott. Geogr. Mag.* **85,** 34–46.

1970. Landforms of deglaciation in the Cairngorm Mountains of Scotland. *Trans. Inst. Br. Geogr.* **51,** 201–19.

1977. Did glaciers form in the Cairngorms in the 17th–19th centuries? *Cairngorm Club Journal* **18,** 189–201.

1980. The Loch Lomond Advance in the Cairngorms—a reply to J. B. Sissons. *Scott. Geogr. Mag.* **96,** 18–19.

SUGDEN, D. E. & CLAPPERTON, C. M. 1975. The deglaciation of Upper Deeside and the Cairngorm Mountains. In Gemmell, A. M. D. (ed.) *Quaternary studies in north-east Scotland,* 30–38. Aberdeen.

SUGDEN, D. E. & JOHN, B. S. 1976. *Glaciers and landscape.* London.

SUTHERLAND, D. G. 1980. Problems of radiocarbon dating deposits from newly deglaciated terrain: examples from the Scottish Lateglacial. In Lowe, J. J., Gray, J. M. & Robinson, J. E. (eds.) *Studies in the Lateglacial of north-west Europe,* 139–150. Oxford.

1981. The high level marine shell beds of Scotland and the build up of the last Scottish ice sheet. *Boreas* **10,** 247–54.

SYNGE, F. M. 1966. The relationship of the raised strandlines and main end-moraines on the Isle of Mull, and in the district of Lorn, Scotland. *Proc. Geol. Ass.* **77,** 315–28.

1977. Records of sea-levels during the Late Devensian. *Phil. Trans. R. Soc. Lond.* B. **280,** 211–28.

SYNGE, F. M. & STEPHENS, N. 1966. Late- and Post-glacial shorelines, and ice limits in Argyll and north-east Ulster. *Trans. Inst. Br. Geogr.* **39,** 101–25.

THOMPSON, K. S. R. 1972. The last glaciers in western Perthshire. *Univ. of Edinburgh Ph.D. thesis.* Unpubl.

THOMSON, M. E. & EDEN, R. A. 1977. Quaternary deposits of the central North Sea III. The Quaternary sequence in the west-central North Sea. *Rep. Inst. Geol. Sci. Lond.* 77/12. 18 pp.

TIVY, J. (ed.) 1973. *The organic resources of Scotland.* Edinburgh.

TROELS-SMITH, J. 1956. Neolithic period in Switzerland and Denmark. *Science* (N.Y.) **124,** 876–9.

1960. Ivy, mistletoe and elm: climatic indicators—fodder plants: a contribution to the interpretation of the pollen zone border VII–VIII. *Danm. Geol. Unders.* IV series, **4,** 1–32.

TURNER, J. 1964. Anthropogenic factor in vegetation history. *New Phytol.* **63,** 73–89.

1965. A contribution to the history of forest clearance. *Proc. R. Soc. Lond.* B. **161,** 343–54.

1970. Post-Neolithic disturbance of British vegetation. In Walker, D. & West, R. G. (eds.) *Studies in the vegetational history of the British Isles: essays in honour of Harry Godwin,* 97–116. London.

1981. The iron age. In Simmons, I. G. & Tooley, M. J. (eds.) *The environment in British pre-history,* 125–219. London.

VAN DONK, J. 1976. An ^{18}O record of the Atlantic Ocean for the entire Pleistocene. *Geol. Soc. Am. Mem.* **145,** 147–64.

VAN GEEL, B. & KOLSTRUP, E. 1978. Tentative explanation of the Lateglacial and early

Holocene climatic changes in north-western Europe. *Geol. En. Mijnbouw* **57,** 87–89.

VASARI, Y. 1977. Radiocarbon dating of the Lateglacial and early Flandrian vegetational succession in the Scottish Highlands and the Isle of Skye. In Gray, J. M. & Lowe, J. J. (eds.) *Studies in the Scottish Lateglacial environment,* 163–82. Oxford.

VASARI, Y. & VASARI, A. 1968. Late- and Postglacial macrophytic vegetation in the lochs of northern Scotland. *Acta. Bot. Fenn.* **80,** 1–120.

VON WEYMARN, J. 1974. Coastline development in Lewis and Harris, Outer Hebrides, with particular reference to the effects of glaciation. *Univ. of Aberdeen Ph.D. thesis.* Unpubl.

VON WEYMARN, J. & EDWARDS, K. J. 1973. Interstadial site on the island of Lewis, Scotland. *Nature* **246,** 473–4.

WAGER, D. R. 1953. The extent of glaciation in the island of St. Kilda. *Geol. Mag.* **90,** 177–81.

WALKER, M. J. C. 1975. Lateglacial and early Postglacial environmental history of the central Grampian Highlands, Scotland. *J. Biogeogr.* **2,** 265–84.

WALKER, M. J. C. & LOWE, J. J. 1977. Postglacial environmental history of Rannoch Moor. I. Three pollen diagrams from the Kingshouse area. *J. Biogeogr.* **4,** 333–51.

WATSON, J. W. 1939. Forest or bog: man the deciding factor. *Scott. Geogr. Mag.* **55,** 148–61.

WATTS, W. A. 1980. Regional variation in the response of vegetation to Lateglacial climatic events in Europe. In Lowe, J. J. *et al* (eds.) *Studies in the Lateglacial of north-west Europe,* 1–22. Oxford.

WELIN, E. *et al* 1975. Institute of Geological Sciences radiocarbon dates VI. *Radiocarbon* **17,** 157–9.

WENDLAND, W. M. & BRYSON, R. A. 1974. Dating climatic episodes of the Holocene. *Quat. Res.* **4,** 9–24.

WEST, R. G. 1977. *Pleistocene geology and biology.* London.

WHITTINGTON, G. 1980. Prehistoric activity and its effect on the Scottish landscape. In Parry, M. L. & Slater, T. R. (eds.) *The making of the Scottish countryside,* 23–44. London.

WIGLEY, T. M. L. *et al* 1981. *Climate and history.* Cambridge.

WRIGHT, W. B. 1911. On a Pre-glacial shoreline in the Western Isles of Scotland. *Geol. Mag.* **48,** 97–109.

1928. The raised beaches of the British Isles. *First Rep. of Comm. on Pliocene and Pleistocene terraces (*Int. Geogr. Union), 99–106.

1937. *The Quaternary ice age.* London.

YOUNG, J. 1864. On the former existence of glaciers in the high grounds of the south of Scotland. *Q. J. Geol. Soc. Lond.* **20,** 452–62.

YOUNG, J. A. T. 1969. Variations in till macrofabric over very short distances. *Bull. Geol. Soc. Am.* **80,** 2343–52.

1974. Ice wastage in Glenmore, upper Spey Valley, Inverness-shire. *Scott. J. Geol.* **10,** 147–58.

1975. A reinterpretation of the deglaciation of Abernethy Forest, Inverness-shire. *Scott. J. Geol.* **11,** 193–206.

1977. Glacial geomorphology of the Dulnain Valley, Inverness-shire. *Scott. J. Geol.* **13,** 59–74.

INDEX